Fire
Data Analysis
Handbook

By
Tom McEwen
with
Catherine A. Miller

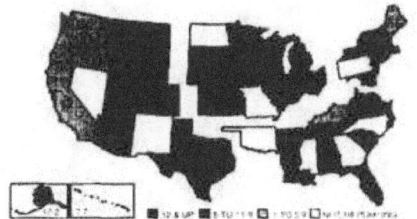

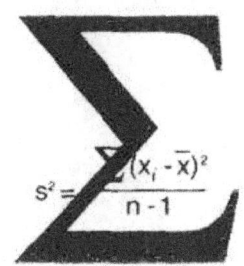

$$s^2 = \frac{\sum (x_i - \bar{x})^2}{n-1}$$

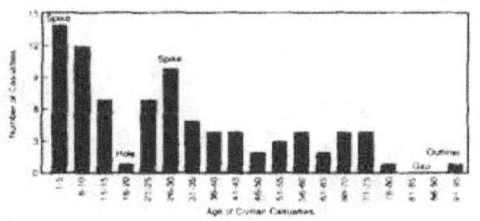

Federal Emergency Management Agency
United States Fire Administration

Foreword

The fire service exists today in an environment constantly inundated with data, but data are often of little use in the everyday, real world in which first responders live and work. This is no accident. By itself, data are of little use to anyone. Information, on the other hand, is very useful indeed. What's the difference? At the recent Olympic games in Korea, a stadium full of people held up individual, multi-colored squares of cardboard to form a giant image or text which could only be recognized from a distance. This is a good analogy for data and information. The individual squares of cardboard are like data They are very numerous and they all look pretty much alike taken by themselves. The big image formed from the organization of thousands of those cards is like information. It is what can be built from many pieces of data. Information then is an organization of data that makes a point about something.

The fire service of today is changing. More and more, it is not fighting tires as much as it is doing EMS, hazmat, inspections, investigations, prevention, and other non-traditional but important tasks which are vital to the community. Balancing limited resources and justifying daily operations and finances in the face of tough economic times is a scenario that every department can relate to. How well a department can do this depends mainly on how well it uses information.

Turning data into information is neither simple nor easy. It requires some knowledge of the tools and techniques used for this purpose. Historically, the fire service has had few of these tools at its disposal and none of them has been designed with the fire service in mind. This book changes that. It was designed solely for the use of the fire service. The examples and problems were developed from fire data collected from departments all over the nation. This book was also designed to be modular in form. Many departments' information needs can be met by studying only the first few chapters. Others with a more statistical bent may want to dig deeper. The point is, it's up to the reader to decide. This book is just another tool, like a pumper or a ladder, to help do the job.

The United States Fire Administration

Table of Contents

Chapter 1
INTRODUCTION

This handbook has a primary objective to describe statistical techniques for analyzing data typically collected in fire departments. Motivation for the handbook comes from the belief that fire departments collect an immense amount of data, but do very little with the data. Think for a minute about the reports you complete on incidents. You probably document the type of situation found, action taken, time of alarm, time of arrival, time completed, number of engines responding, number of personnel responding, and many other items. For fires, the list grows even longer to include area of fire origin, form of heat of ignition, type of material involved, and other related facts. If civilian or fire fighter injuries occur, you complete other reports.

A compelling reason for these reports is a legal requirement for documenting incidents. Victims, insurance companies, lawyers, and many others want copies of reports. Indeed, fire departments maintain files for retrieval of individual reports.

The reports can, however, provide a more beneficial service to fire departments by providing insight into the nature of fires and injuries in your jurisdiction. Basic information is probably already available. Someone usually tracks the number of fires handled last year, the number of fire-related injuries, and the number of fire deaths. It is another story, however, if you ask more probing questions:

- How many fires took place on Sundays, Mondays, etc.?
- How many fires took place each hour of the day or month of the year?
- What was the average response time to fires? How much did response times vary by fire station areas?
- What was the average time spent at the fire scene and how much did the average vary by type of fire?

In a nutshell, this handbook describes statistical techniques to turn data into information for answering these questions and many others. The techniques range from simple to complex. For example, the next two chapters describe how to develop charts to provide more effective presentations about fire problems. These charts may be beneficial to city or county officials on the activities and needs of your fire department. Chapter 4 tells how to compute simple statistics, such as means, medians, and modes. In Chapter 5, we discuss tables and how to calculate different percentages from tables. Other chapters present more sophisticated techniques, such as correlation, regression, loglinear analysis, and queueing theory. These are all techniques which can tell you more about the nature of fires and injuries.

One way to become more comfortable with analysis is to work with real

2 data. For this handbook, data were obtained from fire departments in several metropolitan areas, including Seattle, Washington; Chicago, Illinois; Detroit, Michigan; Jacksonville, Florida; Los Angeles County, California; Monroe County, New York; Boston, Massachusetts; and Dallas, 'Texas. Data on medical emergencies were obtained from the fire department in Prince William County, Virginia, which has completed detailed reports on its responses since 1989. By working with real data, it should be easier for you to understand different techniques.

Why Data Analysis?

There still may be a question in your mind as to why we should go to all this trouble to analyze data. Many decisions do not require analysis, such as decisions on personnel, grievance proceedings, promotions, and even decisions on how to handle a fire. It is certainly true that fire departments can continue to operate in the same way they always have without doing a lot of analysis.

On the other hand, we can give three good reasons for looking more closely at your data: (I) to gain insights into fire problems, (2) to improve resource allocation for combatting fires, and (3) to identify training needs. Probably the most compelling is that analysis gives insight into your fire problems which, in turn, can impact operations in your department. You may find, for example, that the average time to fires in an area is 6 minutes, compared to less than 2 minutes overall. This result may assist you in requests for more equipment, more personnel, or justifing another fire station.

As an example of improved resource allocation, statistical analysis of emergency medical calls can determine the impact of providing another paramedic unit in the field. Increasing the number of EMS units from 4 units to 5 units may, for example, *decrease* average response times from 5 minutes to 3 minutes-a change that may save lives. Chapter 9 describes a queueing model for conducting this type of analysis.

Another reason for analysis is to identify training needs. Most training on fire fighting is based on a curriculum which has been in place for many years. It makes sense to see how training matches characteristics of fires in your own jurisdiction. This is not to sap that other training is unimportant, because an exception can always occur. However, knowing more about your fires can improve your training.

In summary, this handbook will help you deal with the volume of data collected on fire incidents. By studying the techniques presented in this handbook, you should be able to improve your skills in collecting data, analyzing data, and presenting results.

The nest section of this chapter describes the reporting system that serves as the basis for data collection on fires and casualties. 'The importance of quality control is also discussed.

National Fire Incident Reporting System

The National Fire Incident Reporting System (NFIRS) began over 15 years ago with the aim of collecting and analyzing data on fires from departments across the country. More than 13,000 fire departments now report their fires and injuries to NFIRS.

Exhibit 1-1 shows the basic fire incident report from NFIRS. Your department may use this incident report, or you may have a modified version of it. In either case, the data collected is the same and covers all the elements of fire incidents. Lines A through I are completed on all incidents to which a fire department responds. These lines include incident number, date, day of week, alarm time, arrival time, time in service, type of situation found, and type of action taken.

Exhibit 1-1 Basic Fire Incident Report-NFIRS

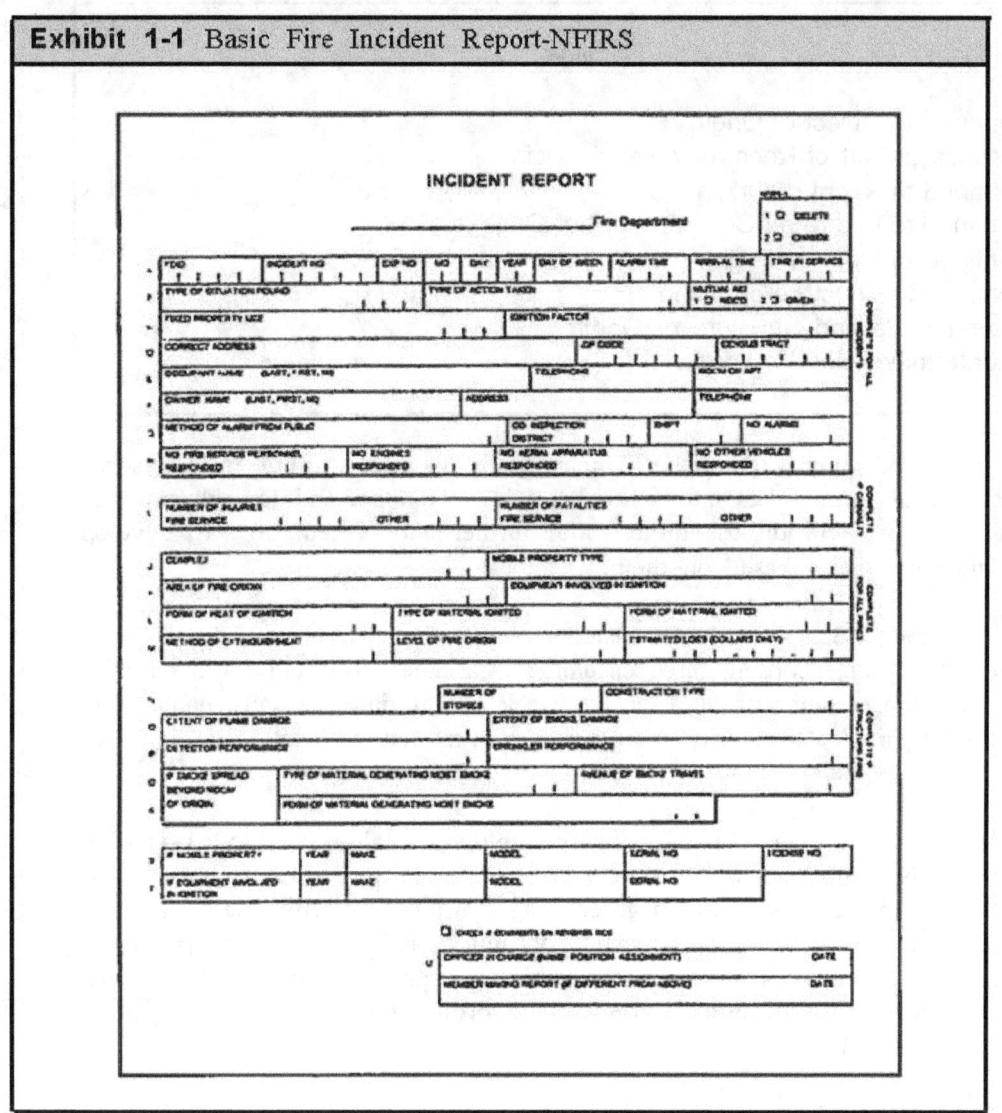

4   Lines J through M are completed on all incidents which are, in fact, fires. They include type of complex, mobile property type, area of fire origin, equipment involved in ignition, from of heat of ignition, dollar loss, and others. Finally, Lines N through R are completed on all structure fires. They include number of stories, construction type, extent of flame damage, detector performance, sprinkler performance, type of material generating the most smoke, avenue of smoke travel, and form of material generating the most smoke. The last two lines (Lines S and T) are for mobile property and equipment involved.

Most items are recorded as numeric codes with the NFPA 901 Codes[1] serving as the coding source. For example, Exhibit 1-2 shows the codes for extent of flame damage (Line 0) of the incident form.

Exhibit 1-2 Extent of Flame Damage	
Categories for Extent of Flame Damage	Code
Confined to Object of Origin	1
Confined to Part of Room or Area of Origin	2
Confined to Room of Origin	3
Confined to Fire-rated Compartment of Origin	4
Confined to Floor of Origin	5
Confined to Structure of Origin	6
Extended Beyond Structure of Origin	7
Undetermined/Not Reported	0

The code numbers have no meaning by themselves, hut instead serve as a way of getting data into a computer in a compact and logical form. In Chapter 5, we will discuss these codes further and present how to develop and interpret tables based on them.

Exhibit 1-3 is the civilian casualty report which is part of NFIRS. Each form allows for recording three casualties. Variables collected for a casualty include incident number, date, day of week, alarm time, casualty name, age, time of injury, sex, severity, familiarity with structure, location at ignition, and condition before injury.

The last form in NFIRS is for fire tighter casualties (Exhibit 1-4). Fire departments complete this form whenever an injury to a fire fighter occurs. This form includes the fire fighter's age, injury severity, part of body injured, activity prior to injury, cause of injury, and medical care provided. In addition, it contains a section on protective equipment for the fire fighter (coat, trousers, boots/shoes, helmet, face protection, breathing apparatus, and gloves).

---

1. For more infortmation on these codes. see *NFPA 901, Uniform Coding for Fire Protection 1976* (National Fire Protection Association. Batterymarch Park. Quincy. Massachusetts. 02269).

Exhibit 1-3 Civilian Casualty Report

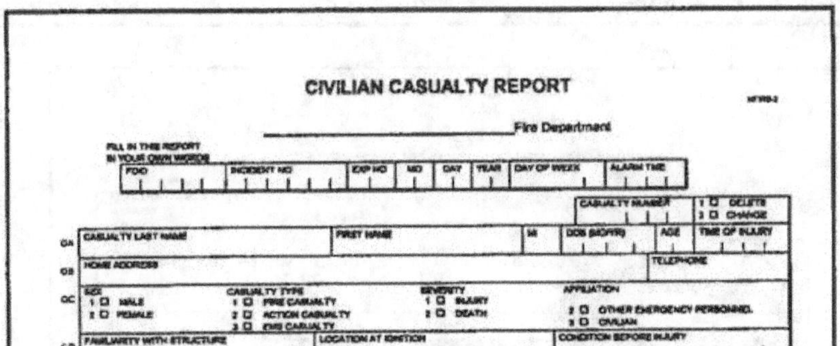

CIVILIAN CASUALTY REPORT

________________ Fire Department

FILL IN THIS REPORT IN YOUR OWN WORDS

FDID | INCIDENT NO | EXP NO | MO | DAY | YEAR | DAY OF WEEK | ALARM TIME

CASUALTY NUMBER | 1 ☐ DELETE 2 ☐ CHANGE

CA CASUALTY LAST NAME | FIRST NAME | MI | DOB (MO/YR) | AGE | TIME OF INJURY

CB HOME ADDRESS | TELEPHONE

CC SEX 1 ☐ MALE 2 ☐ FEMALE | CASUALTY TYPE 1 ☐ FIRE CASUALTY 2 ☐ ACTION CASUALTY 3 ☐ EMS CASUALTY | SEVERITY 1 ☐ INJURY 2 ☐ DEATH | AFFILIATION 2 ☐ OTHER EMERGENCY PERSONNEL 3 ☐ CIVILIAN

CD FAMILIARITY WITH STRUCTURE | LOCATION AT IGNITION | CONDITION BEFORE INJURY

CE CONDITION PREVENTING ESCAPE | ACTIVITY AT TIME OF INJURY | CAUSE OF INJURY

CF NATURE OF INJURY | PART OF BODY INJURED | DISPOSITION

☐ SEE REMARKS ON BACK | ☐ SEE ADDITIONAL REPORT

CASUALTY NUMBER | 1 ☐ DELETE 2 ☐ CHANGE

CA CASUALTY LAST NAME | FIRST NAME | MI | DOB (MO/YR) | AGE | TIME OF INJURY

CB HOME ADDRESS | TELEPHONE

CC SEX 1 ☐ MALE 2 ☐ FEMALE | CASUALTY TYPE 1 ☐ FIRE CASUALTY 2 ☐ ACTION CASUALTY 3 ☐ EMS CASUALTY | SEVERITY 1 ☐ INJURY 2 ☐ DEATH | AFFILIATION 2 ☐ OTHER EMERGENCY PERSONNEL 3 ☐ CIVILIAN

CD FAMILIARITY WITH STRUCTURE | LOCATION AT IGNITION | CONDITION BEFORE INJURY

CE CONDITION PREVENTING ESCAPE | ACTIVITY AT TIME OF INJURY | CAUSE OF INJURY

CF NATURE OF INJURY | PART OF BODY INJURED | DISPOSITION

☐ SEE REMARKS ON BACK | ☐ SEE ADDITIONAL REPORT

CASUALTY NUMBER | 1 ☐ DELETE 2 ☐ CHANGE

CA CASUALTY LAST NAME | FIRST NAME | MI | DOB (MO/YR) | AGE | TIME OF INJURY

CB HOME ADDRESS | TELEPHONE

CC SEX 1 ☐ MALE 2 ☐ FEMALE | CASUALTY TYPE 1 ☐ FIRE CASUALTY 2 ☐ ACTION CASUALTY 3 ☐ EMS CASUALTY | SEVERITY 1 ☐ INJURY 2 ☐ DEATH | AFFILIATION 2 ☐ OTHER EMERGENCY PERSONNEL 3 ☐ CIVILIAN

CD FAMILIARITY WITH STRUCTURE | LOCATION AT IGNITION | CONDITION BEFORE INJURY

CE CONDITION PREVENTING ESCAPE | ACTIVITY AT TIME OF INJURY | CAUSE OF INJURY

CF NATURE OF INJURY | PART OF BODY INJURED | DISPOSITION

☐ SEE REMARKS ON BACK | ☐ SEE ADDITIONAL REPORT

OFFICER IN CHARGE (NAME, POSITION, ASSIGNMENT) | DATE

MEMBER MAKING REPORT (IF DIFFERENT FROM ABOVE) | DATE

The state fire marshall's office in each NFIRS state has responsibility for collecting data from its tire departments. They usually collect data in two ways. One way is that fire departments without any data processing capability send their written reports to the fire marshall's office. The office then takes responsibility for keying reports into a computer system. Local departments with data processing capabilities may send their data on micro-computer diskettes or magnetic tapes. In either case, the state fire marshall's offtce merges all reports into a database.

The state offices have another important responsibility. They create tapes of all fire records (incidents and casualties) and send the tapes on a quarterly basis to the Federal Emergency Management Agency in Washington, D.C. From these tapes, a national database on fires is created each year. The national database for 1990, for example, contains

6 over 941,000 fire incident records and over 18,000 civilian casualty records.

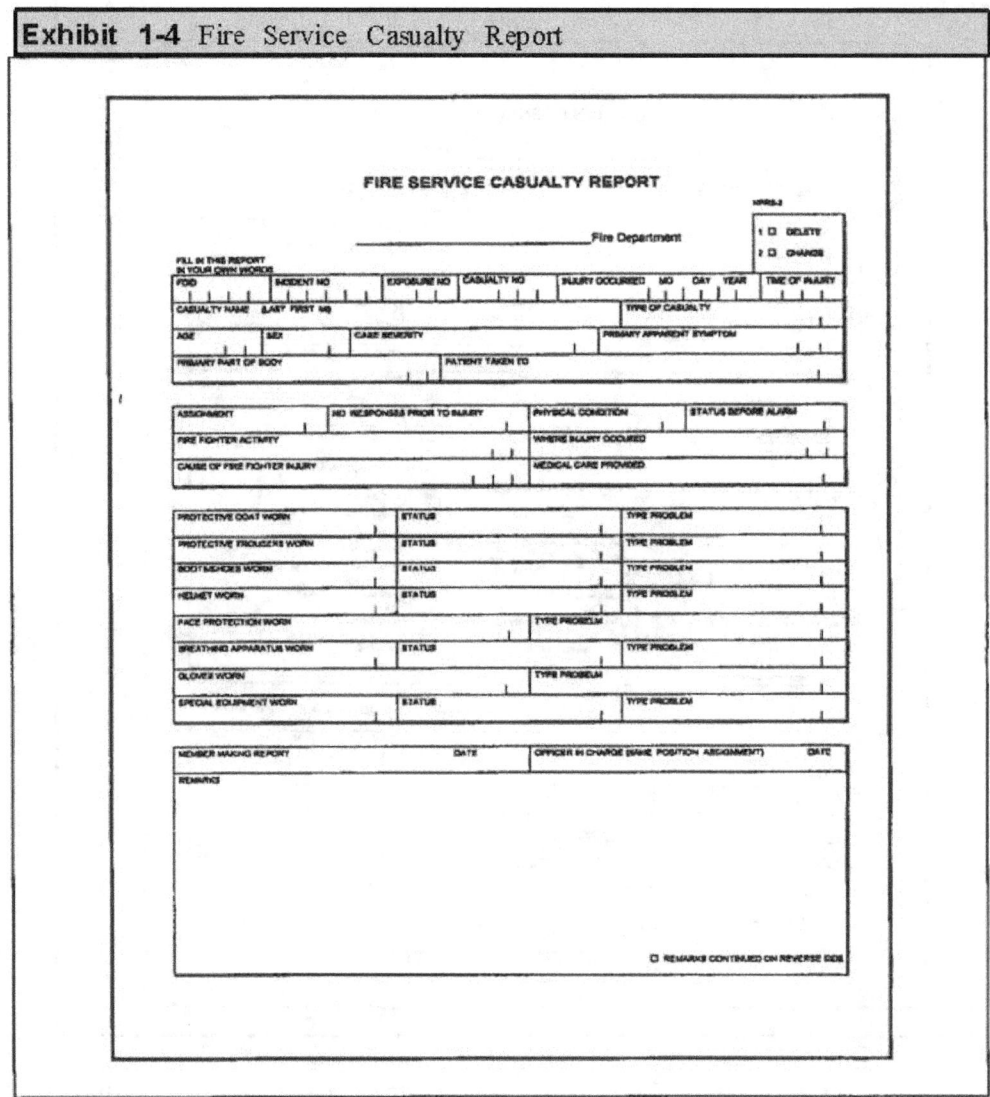

Exhibit 1-4 Fire Service Casualty Report

From a national perspective, this database is vitally important. It allows us to create a picture of different types of fires across the country. The national database has "strength in numbers." Your department, for example, may have only a few fires, if any, where curtains were ignited (Code 36 for form of material ignited). Nationwide, however, there were 2,395 fires of this type during 1990. These fires can be analyzed to draw conclusions about their causes, extent of damages, dollar losses, and other factors.

Data Entry and Data Quality

An assumption throughout the handbook is that data on fire incidents and casualties have been entered into a computer and are available for analysis. While manual analysis is certainly possible, it is usually avoided because the tedious calculations quickly overwhelm our ability to perform analysis in any meaningful manner. The advantage of a computer is that it processes data quickly and accurately.

An immediate problem is how to get data into a computer in the first place. If you are in a large department, this may not be a problem because you probably have a data processing section to enter data. The section may be within the fire department or somewhere else in your government structure. In either case, this section enters NFIRS reports into a computer. Smaller departments usually depend on microcomputers for data collection and analysis. In fact, there are several microcomputer programs specifically for entering incident and casualty data. These programs are not expensive, ranging from $200 to $600.

One word of caution, however, is that any program you purchase should contain a good error checking routine. Data quality is always a problem, and the old adage "Garbage In, Garbage Out" certainly applies to fire department reports. The entry program should, for example, check each item to make sure a valid code has been entered. Whenever the program encounters an error, you should be given an opportunity to correct the error before the data become part of a database. For example, alarm times obviously cannot have hours greater than 23 and minutes greater than 59. An entry program should check hours and minutes for valid numbers, and allow you to make corrections immediately. Similarly, extent of flame damage cannot be coded with an 8 or 9 because these numbers are undefined for this variable (see Exhibit 1-2). Of course, alphabetic characters are also invalid codes.

There is a difference, however, between an "invalid" code and a "wrong" code. By an invalid code, we mean that the code is not on the list of possible codes. A different situation occurs when you enter a "2" instead of a "3" for Extent of Flame Damage. Then you have entered the wrong code.

Wrong codes also occur when blanks or zeroes appear. NFIRS allows blanks because the data are not immediately available and will be determined later. Items particularly susceptible to blanks are the following:

- Mobile property type
- Number of alarms
- Number of fire service personnel
- Number of engines
- Number of aerials
- Complex
- Level of origin
- Number of stories
- Detector performance
- Sprinkler performance

A similar situation exists with coding zeroes. A zero usually indicates that something either could not be determined or was not reported. Items

Chapter 1

in which zeroes frequently occur are ignition factor, form of heat of ignition, type of material ignited, and form of material ignited.

The problem with entering blanks and zeroes initially is that a fire department may never have an opportunity to conduct a follow-up for the correct codes. The blanks and zeroes become a permanent part of the computer record.

While we are on the subject of quality, it is worth exploring the consequences of errors. Obviously, invalid and wrong codes can result in wrong conclusions. If only a few errors appear in the data, the impact on conclusions may not be substantial. On the other hand, a review of national data shows the number of fire personnel (Line 11) is blank in over 25 percent of the fire records, and this amount of missing data has a substantial impact on analysis of responses to fires.

As another illustration, consider what happens when errors appear in alarm time and arrival time. One problem, which surfaced in virtually all fire department data we analyzed, is an occasional reversal of two times. Suppose the alarm time is 1023 and the arrival time is 1027, but the times are reversed on the incident report so that the computer record shows 1027 as the alarm time and 1023 as the arrival time. If we reviewed this report carefully, we would undoubtedly catch and correct this problem before entry into the computer. However, most entry programs will accept these two times without realizing the error. The adverse consequences can be seen when the computer calculates response time, which is defined as the elapsed time between alarm time and arrival time. The response time should be 4 minutes (from 1023 to 1027), hut the reversal creates a situation where the computer calculates an elapsed time of 1356 minutes!! It acts as if the alarm time is 1027 of one day and the arrival time is 1023 of the next day.

It takes only a few errors of this type to cause the calculation of average response time to be completely erroneous. If one-half of one percent of your records have reversed times, the overall average response time may be increased from a correct value of 3 minutes to an erroneous average of 10 minutes.

The point is that fire departments need to establish *data quality procedures* if they intend to take full advantage of their data. Data quality procedures mean that blanks and zeroes should be checked to see if better entries can be made. It also means that response times more than 15 minutes, for example. should be checked. On-scene times (between arrival time and time in service) should be checked as well.

Quality control also means that reports arc checked for logical inconsistencies. A simple example is that Lines J through M should always contain data if type of situation found is between 10 and 19 , indicating a fire occurred. Similarly, Lines N through R should always contain data if type of situation found is an 11, indicating a structure fire.

In summary, data entry programs should include code checking routines to identify errors in individual items in the report and errors reflected through inconsistencies between items. Because entry programs cannot be expected to find all errors, fire departments also need data quality procedures to ensure that correct data are entered into their systems.

Statistical Packages for Computers

In this handbook, we present many different types of analysis. Chapter 3, for example, discusses several types of charts, including bar charts, column charts, histograms, line charts, and dot charts. Other chapters show how to calculate statistics, such as means and variances, and how to do more advanced calculations such as chi-square tests, correlations, and regression coefficients.

In the future, you will want to depend on computers with analysis programs to perform these calculations instead of doing them manually. For a good understanding of analysis, you need to know what is involved, but you should not continue in a manual mode. There are several good statistical packages available for both microcomputers and mainframe computers:

Exhibit 1-5

BMDP Statistical Software, Inc.
1440 Sepulveda Boulevard
Los Angeles, California 90025
213-479-7799

NCSS
329 North 1000 East
Kaysville, Utah 84037
801-546-0445

SAS Institute, Inc.
Software Sales Department
SAS Campus Drive
Gary, North Carolina 27513
919-677-8200

SPSS, Inc.
444 N. Michigan Avenue
Chicago, Illinois 60611
312-329-3500

Statistical Sciences, Inc.
1700 Westlake Avenue, N.
Suite 500
Seattle, Washington 98109
206-283-8802

STSC, Inc.
2115 East Jefferson St.
Rockville, Maryland 20852
301-984-5123

SYSTAT, Inc.
1800 Sherman Avenue
Evanston, Illinois 60201-3793
708-864-5670

If you intend to apply the techniques in this handbook, you should acquire and learn how to use one of these packages.

Books on Data Analysis

You can also expand your knowledge of data analysis with several good textbooks. The following are basic books intended for general audiences:

Misused Statistics: Straight Talk for Twisted Numbers by A. J. Jaffe and Herbert F. Spirer (Marcel Dekker, Inc., New York. N.Y., 1987).

Say It With Figures by Hans Zeisel (Harper & Row Publishers, Inc., New York, N.Y., 1985)

Say It With Charts by Gene Zelazny (Dow Jones-Irwin, Inc., Homewood, Illinois, 1985)

There are numerous statistics books which provide more details on the subjects of this handbook. These books assume more background in algebra and statistics than we have assumed. A sampling follows:

Statistics: The Exploration and Analysis of Data by Jay Devore and Roxy Peck (West Publishing Company, New York. N.Y., 1986)

Statistics by David Freedman, Robert Pisani, and Roger Purves (W.W. Norton & Company, New York, N.Y., 1978)

Beginning Statistics with Data Analysis by Frederick Mosteller, Stephen E. Fienberg, and Robert E.K. Rourke (Addison-Wesley Publishing Company, Reading, Massachusetts, 1983)

Statistics: Concepts and Applications by William C. Schefler (The Benjamin/Cummings Publishing Company, Inc., Menlo Park, California, 1988)

Statistics and Data Analysis: An Introduction by Andrew F. Siegel (John Wiley & Sons, Inc., New York, N.Y., 1988)

One note of caution. All these books are general rather than specific to fire departments. However, concepts are clearly explained and can be applied to analysis of fire data. They also give more details about analysis techniques presented in this handbook.

How to Use This Handbook

Data analysis is not an easy process. It requires careful data collection, attention to details, access to statistical programs, and skills on understsnding results. These are not impossible tasks, but require time and patience on your part for success. Equally important, you need experience. In the long run, you can only develop capabilities in analysis by applying techniques from this handbook on actual data sets.

As a final note, one way of thinking about analysis is to consider a four-stage process as illustrated in Exhibit 1-6.

Exhibit 1-6 Analysis Steps

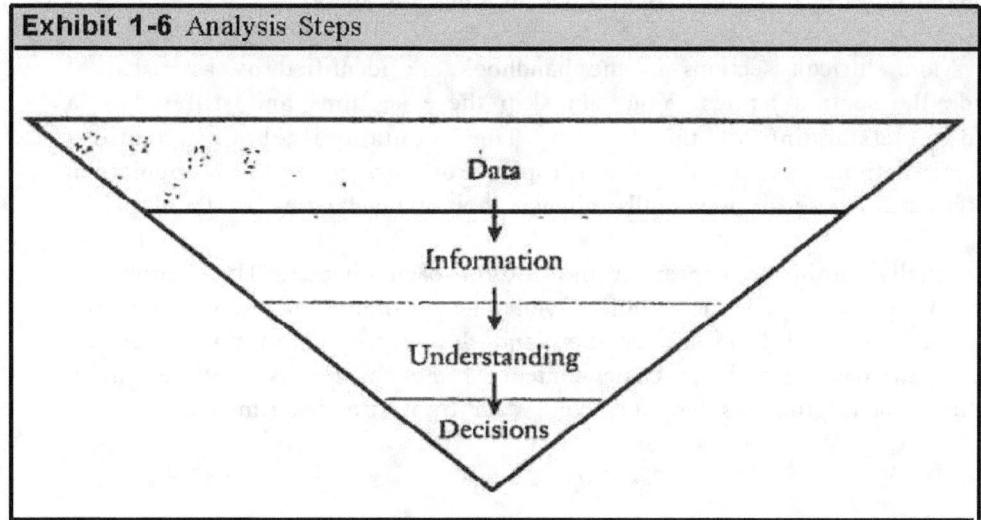

Our ultimate objective is to make better and more informed decisions in fire departments. Data have no utility in a vacuum, and fire reports stay as data if we do nothing. *Analysis turns data into information. We move,* for example, from knowing individual alarm and arrival times to knowing average travel times. Our review of travel times increases our *knowledge* about what is going on with fire incidents which results, in turn, in more informed *decisions* within fire departments.

The remainder of this handbook is organized as follows. We devote Chapters 2 and 3 to descriptions of different types of charts and graphs. Chapter 2 describes histograms, which are probably the easiest and simplest charts to understand. Chapter 3 expands to other types of charts, including bar charts, column charts, pie charts, and dot charts. In Chapter 4, we introduce several basic statistics, including means, medians, modes, and variances. Chapters 5 and 6 discuss analysis of tables, which is particularly important since fire data often comes to us as summaries in the form of tables.

Correlation and regression are the subjects of Chapters 7 and 8. In both chapters, our aim is to present how to perform the calculations associated with these subjects and how to interpret results. Finally, Chapter 9 discusses a modeling technique called queuing theory, which is beneficial for determining the number of emergency medical service units. The number of units depend on the anticipated workload and on predetermined objectives set by a fire department.

In developing these chapters, we recognized that readers will have varying backgrounds and capabilities. The subject material becomes more difficult as you progress through the handhook. 'The first five chapters are easy enough to be read by anyone. More technical subjects, such as regression, are more difficult and require knowledge of basic algebra to understand completely. Even in these chapters, however, we have emphasized understanding results rather than concentrating on theory.

More difficult sections in the handhook are identified by asterisks (*) beside the section names. You can skip these sections and still obtain a good understanding of the subject. They contain algebraic equations for calculations associated with a topic. For persons with mathematical backgrounds, these sections will enhance their understanding of the topic.

Finally, problems appear at the end of each chapter. These problems have two purposes. One is to allow you an opportunity to see if you really understand material from the chapter, and the second purpose is to extend your knowledge beyond the basic content of the chapter. As with examples in the chapter, problems include actual data from fire departments.

Chapter 2
HISTOGRAMS

Data as a Descriptive Tool

"A picture is worth a thousand words" is an old saying which applies to numbers as well as words. The task of reaching conclusions from numbers is formidable, particularly when we are looking for trends and patterns in the data. It is for this reason that we turn our attention to histograms and other charts in this chapter and Chapter 3. These tools will assist us in understanding fire data since the human mind appears to comprehend pictures quicker than words and numbers.

The techniques found in these two chapters include:

Histograms   Column charts   Pie charts   Pictograms

Bar charts   Line charts   Dot charts

This chapter describes histograms while Chapter 3 is devoted to the other techniques. With these graphical aids, we can answer several basic questions. When are fires most likely to occur? What are the primary causes of residential fires? vehicle fires? How many civilian injuries occurred last year by month? What are the ages of civilian casualties? What percent of fire incidents have travel times less than 4 minutes? How many structure fires resulted in dollar losses greater than $50,000 last year?

A **histogram** is a column graph where the height of the columns indicate the relative numbers or frequencies or values of a variable. The following examples show how to organize and display fire data into histograms.

Example 1. One of the most fundamental ways to describe the fire problem is to show how fires are distributed by month, day of week, and hour of day. For example, Exhibit 2-1 shows a **frequency list** of fires by hour of day for Boston, Massachusetts for 1988. A list or array of numbers such as this exhibit is almost always the starting point for a descriptive analysis, hut the numbers by themselves arc not very useful. It is difficult to get a "feel" for what is happening by scanning a list of numbers.

To grasp what the numbers say in Exhibit 2-1, we can develop a frequency histogram, as shown in Exhibit 2-2. Similarly, Exhibits 2-3 and 2-4 show histograms by day of week and month of the year. Study these exhibits for a few minutes and draw your own conclusions about what they say. Don't dwell on individual numbers, but instead look for patterns. Ask yourself three questions:

- Where are the low points and high points in the histogram?

Chapter 2

- What groups of times (hours, days, or months) have similar frequencies?
- Is there anything in the histogram that runs counter to your experience?

Answers to these questions provide the first insights into your fire data and into conclusions from the data.

Exhibit 2-1 Fires by Hour of Day-Boston-1988			
Time Period	Number	Time Period	Number
Midnight - 1 a.m.	478	Noon - 1p.m.	307
1 a.m. - 2 a.m.	420	1 p.m. - 2 p.m.	316
2 a.m. - 3 a.m.	360	2 p.m. - 3 p.m.	363
3 a.m. - 4 a.m.	273	3 p.m. - 4 p.m.	381
4 a.m. - 5 a.m.	192	4 p.m. - 5 p.m.	417
5 a.m. - 6 a.m.	127	5 p.m. - 6 p.m.	433
6 a.m. - 7 a.m.	122	6 p.m. - 7 p.m.	492
7 a.m. - 8 a.m.	139	7 p.m. - 8 p.m.	514
8 a.m. - 9 a.m.	156	8 p.m. - 9p.m.	540
9 a.m. - 10 a.m.	168	9 p.m. - 10 p.m.	622
10 a.m. - 11 a.m.	206	10 p.m. - 11 p.m.	510
11 a.m. - Noon	242	11 p.m. - Midnight	547

It is difficult to get a "feel" for what is happening by scanning a list of numbers.

While these histograms suggest several conclusions, the key ones are:

1. The peak time period for fires in Boston is from 8 pm. to midnight with the hour from 9 p.m. to 10 p.m. having more fires than any other hour.
2. The lowest time period for fires is from 5a.m. to 9 a.m.
3. Weekends are the busiest times for fires while Thursdays are the least busiest days.
4. June and July have more fires than any of the other months while Januay and February have the fewest.

> A **histogram** is a column graph where the height of the bars indicate the relative numbers or frequencies for values of a variable. The values may be numeric, such as travel times, or non-numeric, such as days of the week.

With these histograms we begin to see a picture of the fire problem in Boston. Histograms allow for an easy descriptive and analytical procedure without having to think too much about the numbers themselves. Graphical displays should always strive to convey an immediate message describing a particular aspect of the data.

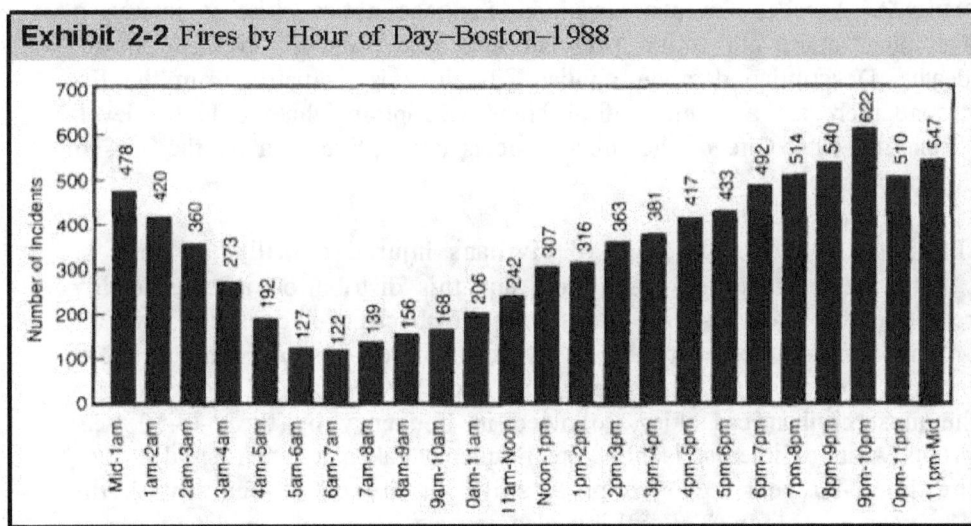

Exhibit 2-2 Fires by Hour of Day–Boston–1988

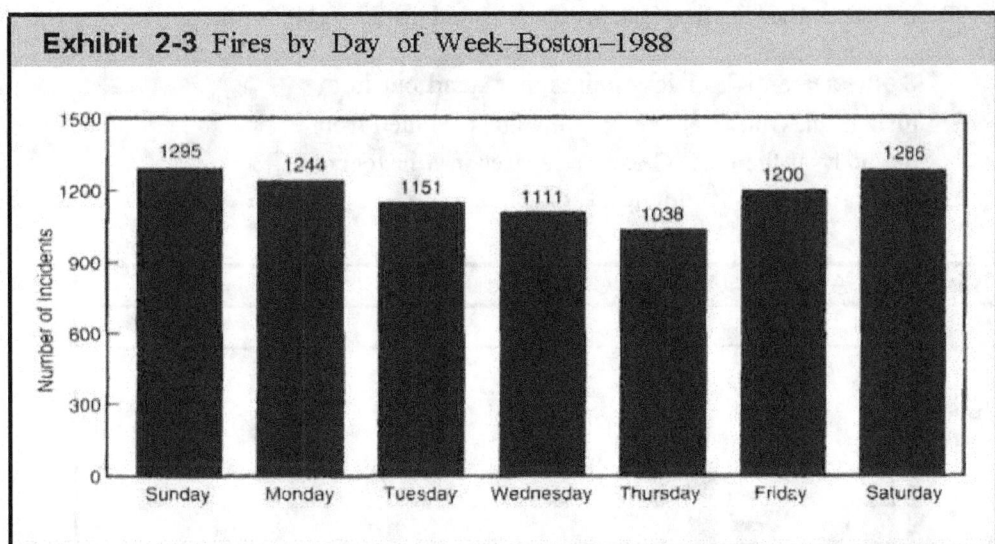

Exhibit 2-3 Fires by Day of Week–Boston–1988

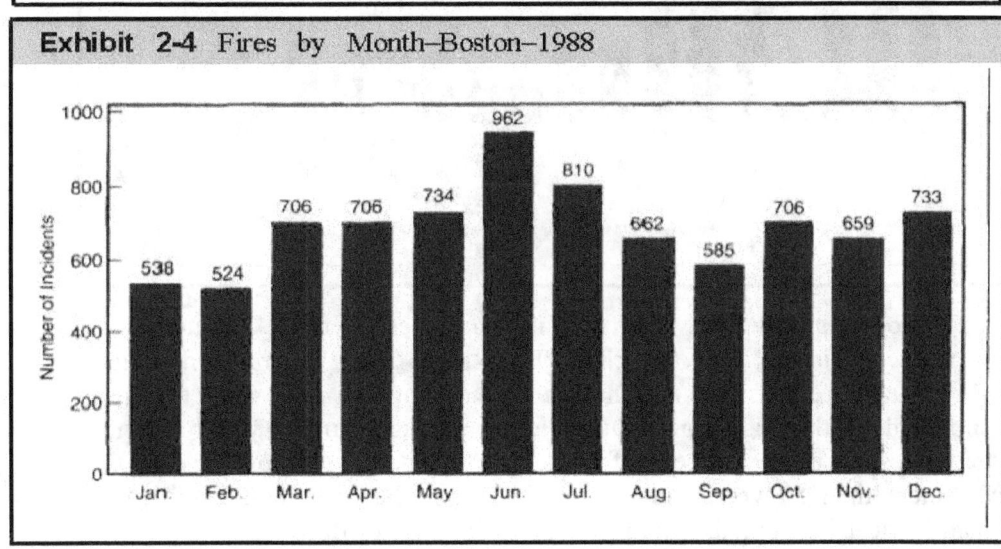

Exhibit 2-4 Fires by Month–Boston–1988

Chapter 2

Example 2. Ages of Civilian Casulties. Suppose a fire chief is interested in developing a fire prevention program aimed at reducing civilian injuries and deaths. Descriptive data on civilian casualtics is available from the fire reports and there arc a number of different descriptions that could be developed from the data. One of the most basic is descriptive data on the ages of civilian casualties.

Exhibit 2-5 shows the ages of civilians injured or killed in fires in Jersey City, New Jersey for 1988. Note that this distribution is considerably different from the previous histograms primarily because it does not have the same "smoothness." However, the five-year age groups show some interesting patterns. For example, the age group under five years of age accounts for the most civilian casualties, followed in frequency by the 6-to-10 year age group. Also of interest is how the frequency takes a rather sudden drop for the 16-to-20 year age group. A spike in the data occurs with the 26-to-30 year age group. The exhibit also reveals a gap in the data for ages 81-to-90 and an outlier in the last group from 91-to-95 years of age.

> **Spikes** are high or low points that stand out in a histogram. Outliers are extreme values isolated from the body of the data. **Gaps** are spaces in a histogram reflecting low frequencies of data.

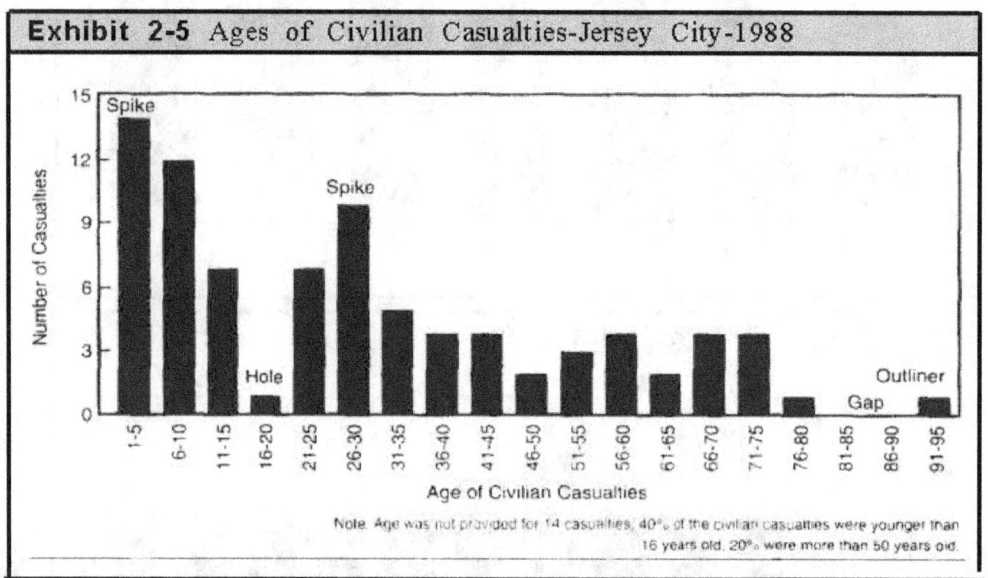

Exhibit 2-5 Ages of Civilian Casualties-Jersey City-1988

Note: Age was not provided for 14 casualties. 40% of the civilian casualties were younger than 16 years old. 20% were more than 50 years old.

In histograms and other charts, it is sometimes useful to include comments and conclusions with the chart. In Exhibit 2-5, we provided a note that 14 casualty records did not include age information and were therefore not included in the histogram. Other notes provide summary information on the data such as the percent of casualties under 16 years of age and the percent more than 50 years old. Anyone studying the histogram could reach the same conclusion, but the summary saves time and effort.

Exhibit 2-6 Travel Times–Seattle–1988

Travel Time	Frequency
Less than 1 minute	106
1 to 2 minutes	85
2 to 3 minutes	481
3 to 4 minutes	1,019
4 to 5 minutes	814
5 to 6 minutes	415
6 to 7 minutes	185
7 to 8 minutes	93
8 to 9 minutes	31
9 to 10 minutes	24
10 to 11 minutes	9
11 to 12 minutes	6
12 to 13 minutes	1
13 to 14 minutes	3
14 to 15 minutes	3
15 to 16 minutes	0
16 to 17 minutes	0
17 to 18 minutes	1
18 to 19 minutes	0
19 to 20 minutes	3
Total Fire Calls	3,279

The data for this example included five other calls with travel times of 26 minutes, 47 minutes, 64 minutes, 683 minutes, and 794 minutes respectively. For purposes of presentation, we have assumed that these records are in error. Either the alarm time or the arrival time has probably been incorrectly recorded. This is obviously the case for the last two average times and the first three times are suspect because they greatly exceed the other times in the distribution. In practice, these times should be reviewed and corrected if necessary.

Example 3. Travel Times to Fires. Travel times to tires are one of the most important data sets to study in fire departments. Many fire departments have objectives for average travel times to fires and try to allocate personnel to achieve these travel times. Exhibit 2-6 shows a frequency distribution for travel times to fires in Seattle, Washington in 1988.

Notice in this example that the times are clustered at the low end of the distribution as we would expect since travel times to fires are generally low for most fire departments.

Exhibit 2-7 provides a frequency histogram for this distribution. In this exhibit, we have combined the last few points into a category of 10 minutes or more. A histogram with the shape in this exhibit is sometimes called **skewed to the right** or **skewed toward high values.** What we mean by these terms is that the distribution is not symmetric but instead has a single peak on the left side of the distribution with a long tail toward the right. In fire departments, data on on-scene time (from time of arrival to time back in service) and data on dollar losses at fires also reflect skewness to the right.

Chapter 2

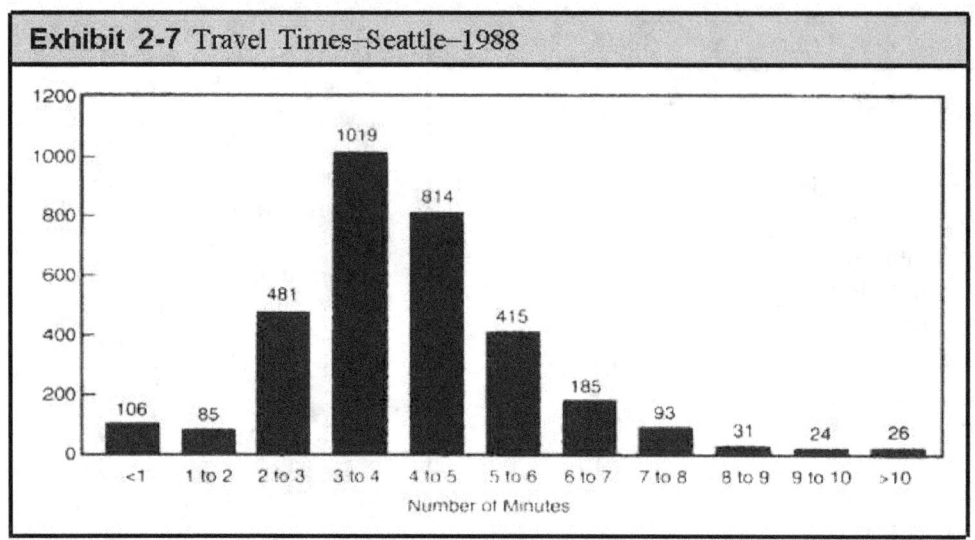

Exhibit 2-7 Travel Times–Seattle–1988

Developing a Histogram

Making a histogram is relatively straightforward:

1. Choose the number of groups for classifying the data. You should usually, have between 5 and 10 groups, hut there are exceptions such as histograms by hour of day. Sometimes the groups are natural, as in our exhibits by day of week and month. With other data, you will have to develop appropriate intervals for the data, as we did with the ages of civilian casualties in Exhibit 2-5.
2. Determine the number of events (fires, casualties, etc.) for each of your groups.
3. For data such as ages and travel times, you usually need to define intervals. For these intervals, you should choose convenient whole numbers. That is, avoid fractions in the groups and always make the intervals the same width. In Exhibit 2-5 we used intervals of five years for grouping the data. Data such as day of week do not require this step since their intervals are naturally defined.
4. Determine the number of observations in each group. Statistical packages are particularly, useful for this step since they always include routines for tabulating data.
5. Choose appropriate scales for each axsis to accommodate the data.
6. Display the frequencies with vertical bars.

Do not expect to get a histogram, or any other type of chart, exactly right on the first try. You may have to try several times before you are satistied with the look of the histogram.

The hisgrams presented in the previous section offer good example of different characteristics for describing the data. Mosteller, et. al. (1983) offer the following definitions of features you should try to find in histograms:

1. **Peaks and valleys.** The peaks and valleys in a histogram indicate the values that appear most frequently (peaks) or least frequently (valleys). Exhibit 2-2 shows clear peaks and valleys for incidents by hour of day.
2. **Spikes and holes.** These are high and low points that stand out in the histogram. In Exhibit 2-5, for example, there is a spike for the 26-to-30 year age group, and a hole for the 16-to-20 year age group.
3. **Outliers.** Extreme values arc sometimes called outliers and are points that are isolated from the body of the data. In Exhibit 2-5, there is one outlier in the 91-to-95 year age group.
4. **Gaps.** Spaces may reflect important aspects of a histogram. In Exhibit 2-5, there is a gap in the 81-to-90 year age group.
5. **Symmetry.** Sometimes a histogram will he balanced along a central value. When this happens, the histogram is easier to interpret. The central value is both the average for the distribution and the median (half the data points will be below this value and half will he above).

Cumulative Frequencies

Two other types of distributions which will be important in later chapters are the **cumulative frequency** and the **cumulative percentage frequency**. A cumulative frequency is the number of data points that are less than or equal to a given value. A cumulative percentage frequency converts the cumulative frequencies into percentages.

Example 4. With the data in Exhibit 2-6, we can calculate the cumulative frequency and cumulative percentages for the travel time data from Seattle, Washington.

Exhibit 2-8 Cumulative Travel Times–Seattle–1988

Travel Time	Frequency	Cumulative Frequency	Cumulative Percent
Less than 1 minute	106	106	3.2
1 to 2 minutes	85	191	5.8
2 to 3 minutes	481	672	20.5
3 to 4 minutes	1,019	1,691	51.6
4 to 5 minutes	814	2,505	76.4
5 to 6 minutes	415	2,920	89.1
6 to 7 minutes	185	3,105	94.7
7 to 8 minutes	93	9,198	97.5
8 to 9 minutes	31	3,229	98.5
9 to 10 minutes	24	3,253	99.2
10 or more minutes	26	3,279	100.0
Total		3,279	100.0

A cumulative frequency is the number of data points that are less than or equal to a given value. A cumulative percentage frequency converts the frequencies into percentages.

Chapter 2

The first entry under the "Cumulative Frequency" column is 106, which is the same as in the "Frequency" column. The second entry shows 191, which is 106 + 85, the sum of the first two entries in the "Frequency" column. By adding these two number, we can say that 191 incidents have travel times less than 2 minutes. The next entry is 672 (106+85+481) and means that 672 incidents have travel times less than 3 minutes. The cumulative frequencies continue in this manner with the last entry in the column always equal to the total number of incidents in our analysis.

The last column, labeled "Cumulative Percent" merely converts the cumulative frequencies into percentages. This step is accomplished by dividing each cumulative frequency), by 3,279, which is the total number of incidents. The column shows that 3.2 percent of the incidents have travel times less than 1 minute, 5.8 percent less than 2 minutes, 20.5 percent less than 3 minutes, etc.

> A **cumulative frequency** is the frequency of data less than or equal to a group. A **cumulative percent** is the cumulative frequency divided by the total number of events.

In general, cumulative percentages describe data in "more than" and "less than" terms. We can conclude, for example, that about half the calls have travel times of less than 4 minutes and about 95 percent have travel times less than 7 minutes. Travel times exceed 9 minutes in only about one percent of the calls.

Summary

A list of numbers is frequently the starting point for analysis. If the question of interest is for specific information, then the list of numbers serves the purpose. For example, Exhibit 2-1 is useful if we are asked about exactly how many fires occurred between 2 a.m and 3 a.m., or if we want to know the exact difference between the busiest hour and the least busiest hour. On the other hand, Exhibit 2-1 is not very useful for determining, for example, the six busiest hours of the day.

Histograms provide a much better method for getting the feel of a list of numbers and answering several questions about relationships. The patterns in a histogram are especially important. For example, high frequencies and low frequencies are usually important to note. Trends indicated by) spikes, outliers, and paps in a histogram are also important.

Chapter 2
PROBLEMS

1. With the data in Exhibit 2-1, determine the number of fires by four-hour periods (Midnight to 4 a.m., 4 a.m. to 8 a.m., etc.). Develop a histogram for these four hour periods. What are the advantages and disadvantages of this histogram compared to Exhibit 2-2?

2. What do Exhibits 2-2, 2-3, and 2-4 tell us about when we should schedule firefighters if we want to match fire workload with personnel?

3. The following figures and percentages are for the Boston fires occurring in 1990. Compare these distributions to Exhibits 2-2, 2-3, and 2-4. Note first that there is a substantial reduction in the number of fires from 8,325 fires in 1988 to 6,479 fires in 1990. The comparisons you make should determine whether the distribution of fires has changed from 1988 to 1990. That is, are the busy hours during 1990 the same as the busy hours for 1988?

Boston, 1990: Hour of Day

Time Period	Number	Time Period	Number
Midnight - 1 a.m.	386	Noon - 1 p.m.	213
1 a.m. - 2 a.m.	287	1 p.m. - 2 p.m.	265
2 a.m. - 3 a.m.	210	2 p.m. - 3 p.m.	293
3 a.m. - 4 a.m.	194	3 p.m. - 4 p.m.	295
4 a.m. - 5 a.m.	146	4 p.m. - 5 p.m.	354
5 a.m. - 6 a.m.	95	5 p.m. - 6 p.m.	380
6 a.m. - 7 a.m.	78	6 p.m. - 7 p.m.	384
7 a.m. - 8 a.m.	126	7 p.m. - 8 p.m.	432
8 a.m. - 9 a.m.	141	8 p.m. - 9 p.m.	498
9 a.m. -10 a.m.	138	9 p.m. - 10 p.m.	492
10 a.m. -11 a.m.	156	10 p.m. - 11 p.m.	394
11 a.m. - Noon	183	11 p.m. - Midnight	339

Boston, 1990: Day of Week

Day	Number	Percent
Sunday	965	14.9
Monday	885	13.7
Tuesday	960	14.8
Wednesday	912	14.1
Thursday	906	14.0
Friday	944	14.6
Saturday	907	14.0
Total	6,479	100.0

Boston, 1990: Month

Month	Number	Month	Number
January	508	July	798
February	342	August	493
March	529	September	509
April	548	October	436
May	529	November	580
June	702	December	505

4. Answer the following questions based on the data on civilian casualties from Jersey City for 1988:

Jersey City, New Jersey, 1988: Ages of Civilian Casualties

Age Group	Number	Age Group	Number
1-5	14	51-55	3
6-10	12	56-60	4
11-15	7	61-65	2
16-20		66-70	3
21-25	7	71-75	3
26-30	10	76-80	1
31-35	5	81-85	0
36-40	4	86-90	0
41-45	4	91-95	1
46-50	2		

a. Develop a cumulative frequency distribution and cumulative percentage distribution of the data:

b. What percentage of civilians less than 16 years of age were injured or killed in fires?

c. What percentage of civilians were more than 50 years old?

5. The following data are from the national NFIRS system which was discussed in Chapter 1. The data show the distribution of ages, in five year increments, for 2,280 civilian casualties in 1988. Civilian casualties include persons other than firefighters injured or killed during fires. Develop a histogram from these figures and compare the results to the data from Jersey City, New Jersey in Exhibit 2-5.

Age	Number	Percent	Age	Number	Percent
< 5 years	256	11.2	5 1 - 55 years	69	3.0
6 - 10 years	100	4.4	56 - 60 years	94	4.1
11 - 15 years	90	3.9	61 - 65 years	105	4.6
16 - 20 years	139	6.1	66 - 70 years	76	3.3
21 - 25 years	231	10.1	71 - 75 years	53	2.3
26 - 30 years	275	12.1	76 - 80 years	51	2.2
31 - 35 years	230	10.1	81 - 85 years	47	2.1
36 - 40 years	179	7.9	86 - 90 years	23	1.0
41 - 45 years	139	6.1	91 - 95 years	7	.3
46 - 50 years	114	5.0	96 - 100 years	2	.1

6. For the civilian casualties in the previous problem, the following data show the number of casualties by hour of day. Develop a plot of these casualties and give reasons why this distribution differs from the distribution of fires by hour of day.

Civilian Casualties, 1988: Hour of Day

Time Period	Number	Time Period	Number
Midnight - 1 a.m.	93	Noon - 1 p.m.	66
1 a.m. - 2 a.m.	125	1 p.m. - 2 p.m.	110
2 a.m. - 3 a.m.	118	2 p.m. - 3 p.m.	94
3 a.m. - 4 a.m.	159	3 pm . - -I p.m.	105
4 a.m. - 5 a.m.	101	4 p.m. - 5 p.m.	92
5 a.m. - 6 a.m.	85	5 p.m. - 6 p.m.	112
6 a.m. - 7 a.m.	62	6 p.m. - 7 p.m.	103
7 a.m. - 8 a.m.	69	7 p.m. - 8 p.m.	62
8 a.m. - 9 a.m.	71	8 p.m. - 9 p.m.	88
9 a.m. - 10 a.m.	107	9 p.m. - 10 p.m.	85
10 a.m. - 11 a.m.	108	10 p.m. -11 p.m.	81
11 a.m. - Noon	86	11 p.m. -Midnight	84

NOTE: The time of the fire was not known in 14 cases so that the total for the above figures is 2,266 civilians.

7. How do peaks, holes, spikes and gaps affect cumulative distributions?

Chapter 3
CHARTS

Introduction

In this chapter we will extend beyond histograms to other types of charts. The point is that histograms are only one of many different ways of portraying data. As an analyst, you must decide which type of chart best reflects the results you want to present. A histogram may serve as the best vehicle, but other types of charts should be considered such as bar charts, line charts, pie charts, dot charts and pictograms. Each of these will be explained in this chapter.

Two questions to bear in mind throughout this chapter are the following:

* What are the main conclusions from your analysis?
* What is the best way to display the conclusions?

As with the previous chapter, several sets of data will be presented in this chapter. You should study each example carefully and draw your own conclusions about the results. You may, in fact, disagree with what we emphasize or you may identify an aspect of the data we overlooked. In either case, think about how you would present your viewpoints in a graphical format to a given audience. The audience may be an internal group of managers, an outside association or group of citizens, or even your city or county council. Even the audience influences the type of chart selected.

The first step is therefore to determine the key results you see from the data. Once you have reached conclusions, you want to select the best type of chart to convey those conclusions. Often you will want to try different charts to determine the best presentation for your audience and your data.

Each of the following sections describes a different type of chart. At the end of the chapter, we present guidelines on selecting a type of chart suitable for different conclusions.

Bar Charts

A **bar chart** is one of the simplest and most effective ways to display data.

In a bar chart, we simply draw a bar for each category of data allowing for a visual comparison of the results. For example, the figures in Exhibit 3-1 on the following page give the ignition factors (from NFPA 901 codes) for the 7,509 structure fires in Chicago, Illinois for 1990.

Chapter 3

Our interest in a list of this type usually centers on how the items compare to each other. What is the leading ignition factor in structure fires? How does misuse compare to mechanical deficiency problems? How big a problem are suspicious fires?

We can determine some results relatively easy from the list of numbers. For example, misuse of heat of ignition is clearly the leading ignition factor followed by suspicious fires and mechanical failures. Operational deficiencies account for less than 10 percent of the total. Design, construction, and installation deficiencies also account for less than 10 percent of the total. However, these results require us to make comparisons mentally with numbers or percentages.

A bar chart overcomes these problems by presenting the data in frequenecy order, as displayed in Exhibit 3-2. The horizontal dimension gives

Exhibit 3-1 Ignition Factors for Structure Fires–Chicago, Illinois–1990

Ignition Factor	Number	Percent
00. Not Reported	496	6.6
10. Incendiary	708	9.4
20. Suspicious	1,077	14.3
30. Misuse of Heat of Ignition	2,273	30.3
40. Misuse of Material Ignited	426	5.7
50. Mechanical Failure	950	12.7
60. Design, Construction, Installation Deficiency	734	9.8
70. Operational Deficiency	620	8.3
80. Natural Conditions	16	.2
90. Other Ignition Factors	209	2.8
Total	7,509	100.0

Exhibit 3-2 Ignition Factors for Structure Fires–Chicago, Illinois–1990

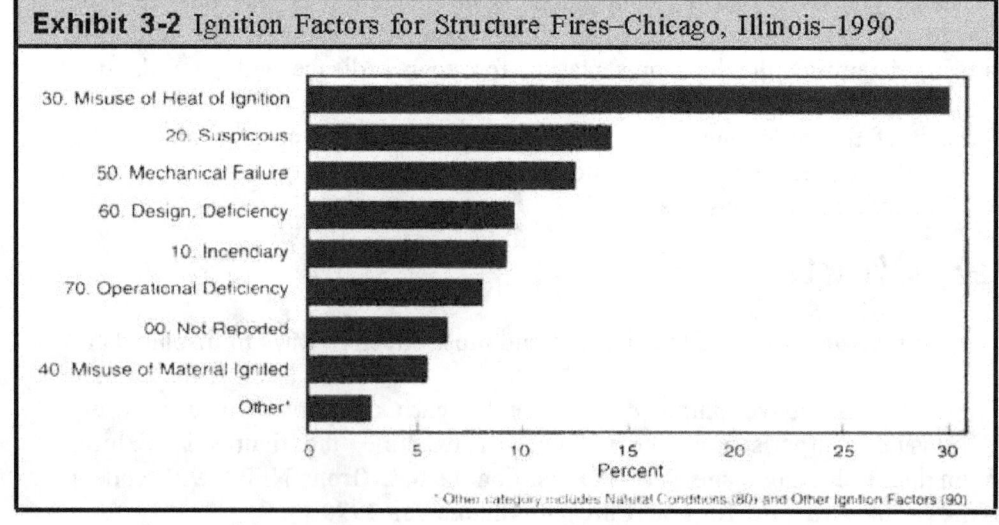

* Other category includes Natural Conditions (80) and Other Ignition Factors (90)

the percent, while the vertical dimension shows the category labels. The bars are presented in numerical order starting with operational deficiencies as the most frequent. Each bar also contains the number of fires for the ignition factor as additional information to a reader.

You should also note that the bottom bar is labeled "Other" with 225 structure fires in the bar. This number is actually the combination of the natural conditions (16 fires) and other ignition factors (209 fires). It is not unusual to combine low frequency categories into an "other" category. However, you should accompany the chart with a table, as we have done here, or add a footnote to the chart indicating the combinations.

As a general rule, the horizontal dimension in a bar chart is numeric, such as percentages or other numbers, while the vertical dimension shows the labels for the items in a category. It is not always necessary to include numbers in each bar, but they are sometimes useful to readers unfamiliar with the data. If you omit numbers in the charts, you should provide the total number of incidents either in the title or as a footnote.

We could have arranged the labels in code number order so that they matched the list in Exhibit 3-1. The emphasis in such a chart would be on the individual categories rather than on their ranking. However, in this case, you might want to highlight the bar of the highest ranking item with stripes or a different color than the rest of the graph. Exhibit 3-3 is an example of this type of chart. The exhibit has horizontal bars indicating the number of non-residential deaths by property type during 1990. As emphasized by the solid bar, more deaths occurred in manufacturing facilities than any other type of property.

A **grouped bar chart** shows two categories in the same chart. In Exhibit 3-4, for example, we display the ignition factors for structure versus vehicle fires. Since there are two categories, we list the items in code number order.

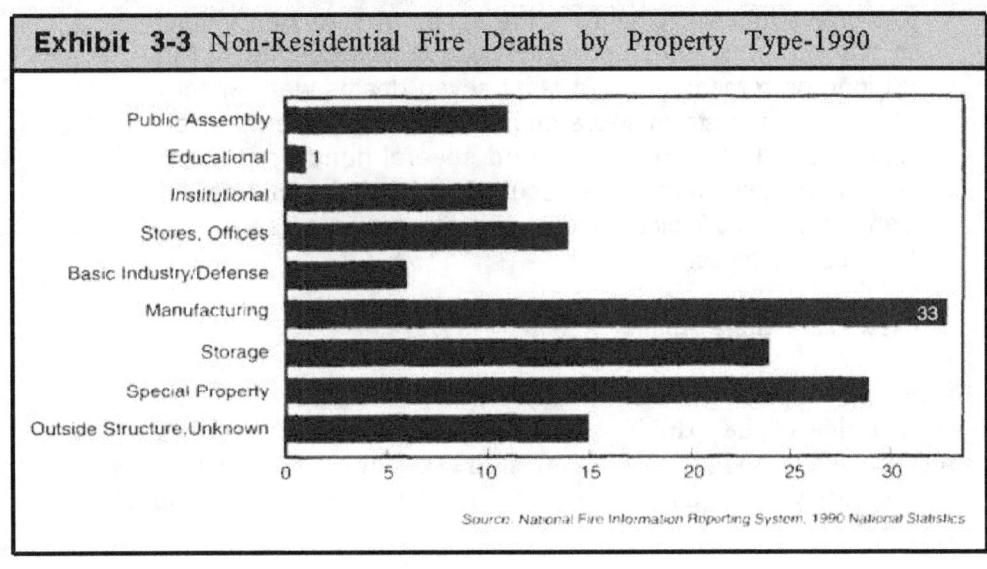

Exhibit 3-3 Non-Residential Fire Deaths by Property Type-1990

Source: National Fire Information Reporting System, 1990 National Statistics

Chapter 3

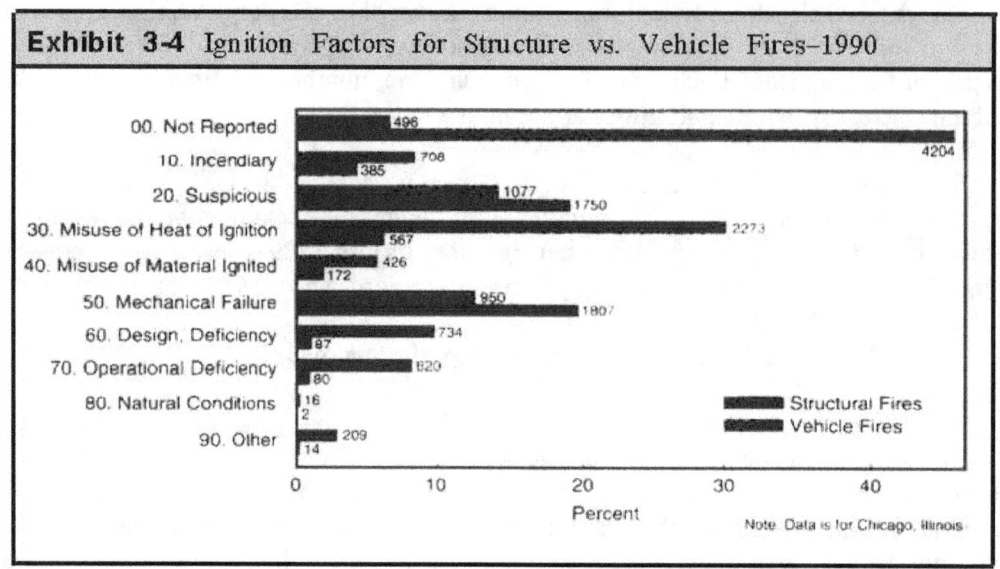

Exhibit 3-4 Ignition Factors for Structure vs. Vehicle Fires–1990

This exhibit shows, for example, that suspicious fires (generally arson fires) are a greater problem with vehicle fires than structure fires. The chart also shows that ignition factors are not reported in more than 4,000 vehicle fires compared to only 496 structure fires. The paired bar chart clearly shows the differences in ignition factors for these two types of fires.

A **paired bar chart** makes item-by-item comparisons. By way of background to an interesting paired bar chart, we provide the following summary of the impact from a snowstorm in New York state in 1987:

> "On Sunday morning, October 4, 1987, the Hudson Valley region of New York State was hit by an unusual, early snow-storm which brought up to two feet of heavy, wet snow. The snow fell in a band from Washington and Saratoga Counties in the north, through the Berkshires and Catskills to the northern part of Westchester County in the south. The sudden impact on fire and other emergency services throughout the region will long be remembered. At least seven deaths were directly attributed to the storm. More than 270 fire departments mobi-lized over 11,000 firefighters and several hundred pumps, generators, saws and related equipment in what would be, for many, a round-the-clock, week-long outpouring of service to their communities."

(New York State Annual Report, 1987).

Exhibit 3-5 displays the number of calls for the 11 most affected counties. The left side of the exhibit shows the average number of calls for past Octobers while the right side shows the October 1987 calls. With this arrangement, we can immediately see the impact of the snowstorm for each county.

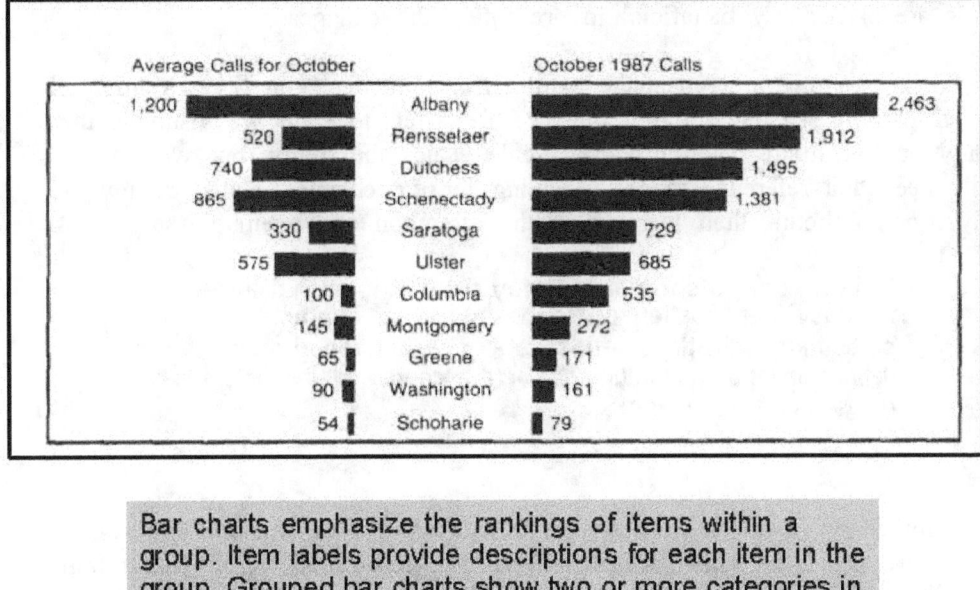

Exhibit 3-5 Calls for October, 1987 vs. Average Calls for Past Octobers

Bar charts emphasize the rankings of items within a group. Item labels provide descriptions for each item in the group. Grouped bar charts show two or more categories in the same chart. Paired bar charts allow for item-by-item comparisons.

Column Charts

We displayed several column charts in Chapter 2. For example, Exhibits 2-2, 2-3, and 2-4 showed Boston fires during 1988 by hour of day, day of week, and month. These are all examples of **time series** presented as column charts.

Column charts of this type are particularly useful in demonstrating

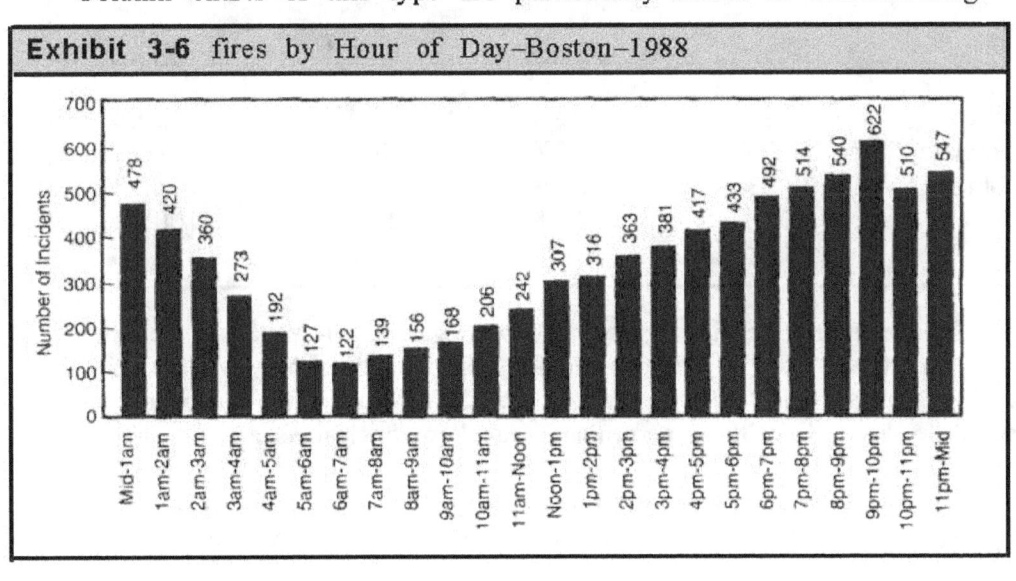

Exhibit 3-6 fires by Hour of Day–Boston–1988

change over time. Where is the series increasing, decreasing, or staying about the same? If our analysis shows changes over time, then column charts are particularly beneficial in presenting the changes.

As an example, we repeat the exhibit from Chapter 2 on fires by hour of day in Exhibit 3-6. By moving our eye from left to right we visualize the change in our mind. The horizontal scale scales the hours, but we do not really need that reference to get a feeling for the changes. Calls are low in early morning hours, then increase in the afternoon and evening hours.

> Column charts show frequency distributions that allow for easy identification of trends and other characteristics, particularly with time series data. The horizontal scale defines the natural groupings for the chart and the columns gives the frequencies.

Another good application of column charts is to show comparisons across several sets of data. Exhibit 3-7 lists fire department activities for four sites divided into fires, rescue calls, and other calls. Comparisons across the sites are not easy because the totals differ so much. Site A has 17,576 calls while the other sites have less than 3,000 calls. A simple way to overcome this problem is to develop percentages.

Exhibit 3-7 Comparison of Fire Department Activities–1988

	Site A	Site B	Site C	Site D
Fires	1,390	170	346	368
Rescue Calls	10,242	636	576	1,668
Other	5,944	576	694	879
Total	17,576	1,382	1,616	2,915
	Site A	Site B	Site C	Site D
Fires	7.9%	12.3%	21.4%	12.6%
Rescue Calls	58.3%	46.0%	35.6%	57.2%
Other	33.8%	41.7%	43.0%	30.2%
Total	100.0%	100.0%	100.0%	100.0%

Note: "Other" calls include Hazardous Conditions, Service calls, Good Intent calls, and False calls

By converting the site figures to percentages, as shown at the bottom of the exhibit, we have a better basis for comparisons. The percentages for each site always add to 100 percent. While there many many conclusions that could he drawn from these percentages, the key conclusions are:

- Fire calls are always the smallest percentage of activity in each site.
- Rescue calls are the predominant type of call in three sites.

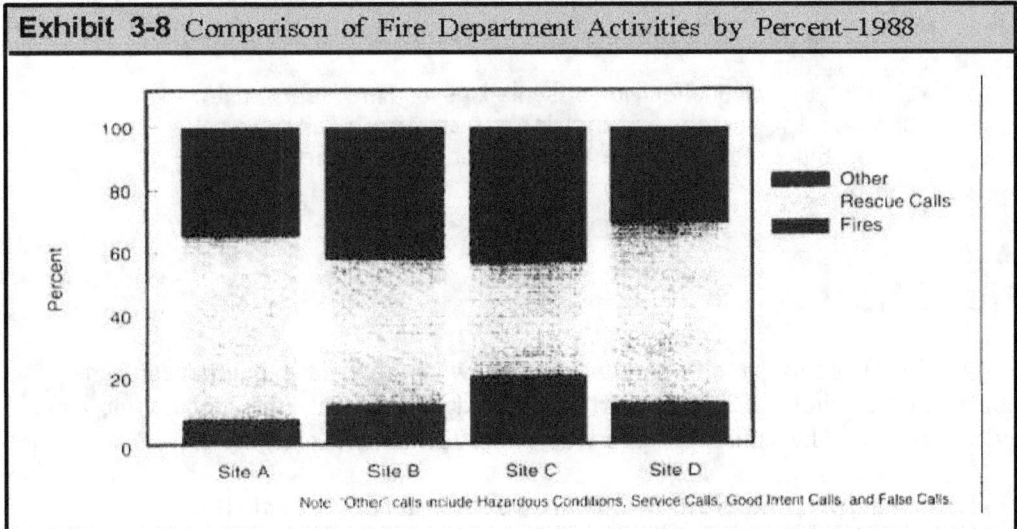

Exhibit 3-8 Comparison of Fire Department Activities by Percent—1988

To display this result, we develop **stacked column charts** as shown in Exhibit 3-8 using the percentages for each activity. The columns all have the same height since they add to 100 percent. Different shadings highlight the amount of activity. The results just discussed should he clear from the exhibit.

Line Charts

Effective presentation of time series data may also be developed from **line charts.** Exhibit 3-9 shows a line chart of fires for Detroit, Michigan, for 1988, 1989, and 1990 by month. The line chart immediately highlights problems during the summer months of 1988 when a substantial number of fires occurred. For 1989 and 1990, these summer months are still among the highest for the years, hut do not begin to approach the problems in 1988. Many statisticians believe that a line

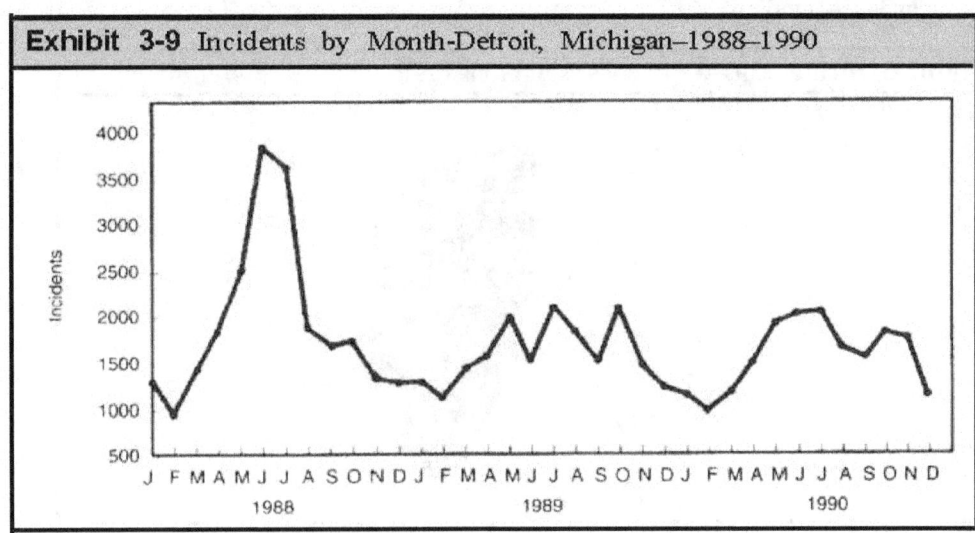

Exhibit 3-9 Incidents by Month-Detroit, Michigan—1988–1990

Chapter 3

chart is the clearest nay for showing increases, decreases, and fluctuations in a time series.

> Line charts give effective presentations of time series data, such as the number of incidents per month for several years. Fluctuations in the data are easily identified by line charts.

Pie Charts

A **pie chart** is an effective way of showing how each component contributes to the whole. In a pie chart, each wedge represents the amount for a given category. The entire pie chart accounts for all the categories.

For example, Exhibit 3-10 shows the activities of the Seattle, Washington Fire Department for 1990 divided into fire calls, false calls, service calls, good intent calls, and other calls. The percentages are inserted in each wedge. Although the percentage numbers are not necessary, they aid in the comparisons of the wedges. The pie chart emphasizes the fact that false calls account for a high percent of incidents in the city. Fire calls and good intent calls account for about the **same** percent **of total** incidents.

In developing pie charts, you should follow the following rules:

- Convert your data to percentages.
- Keep the number of wedges to six or less. If you have more than six, try keeping the most important five and group the rest into a sixth category.
- Position the most important wedge starting at the 12 o'clock position.
- Highlight the most important wedge by coloring it the most intense shading.

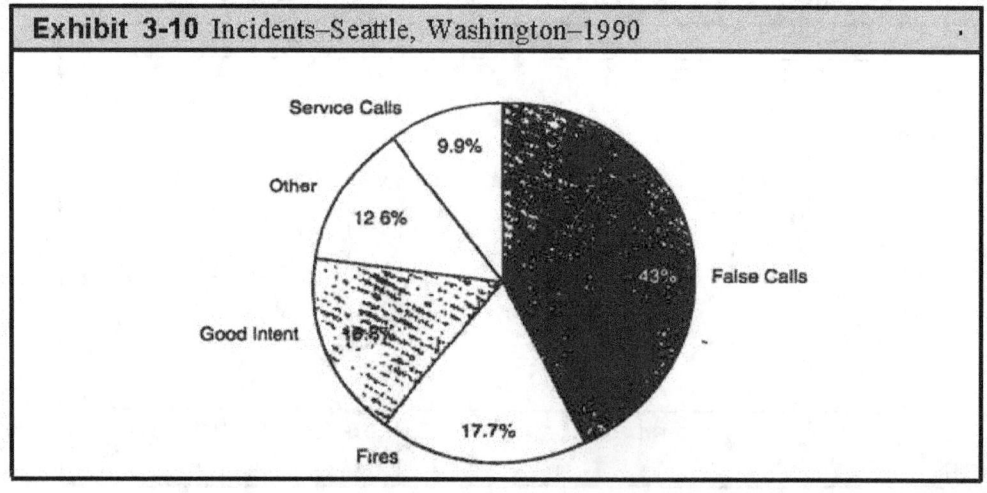

Exhibit 3-10 Incidents–Seattle, Washington–1990

While pie charts are popular, they are probably the least effective way of displaying your results. For example, it may be hard to compare wedges within a pie to determine their ranks. Similarly, it takes time and effort to compare several pie charts because they are separate figures.

> Develop a **pie chart** when the objective is to show how each item contributes to the whole. Pie charts are not effective for comparing several groups of figures.

Dot Charts

Dot charts or **scatter diagrams** emphasize the relationship between two variables. For example, Prince William County, Virginia experienced increases in population and Emergency Medical Services (EMS) calls over the last few years. In 1981, the department responded to 9,538 EMS calls as compared to 12,744 EMS calls in 1991. During these years, the population increased steadily from 152,300 to 223,900. We would expect EMS calls to increase with population, and it is this relationship that we want to depict in a chart.

Exhibit 3-11 is a dot chart for EMS calls versus population for Prince William County for the eleven years from 1981 to 1991. Population is along the horizontal axis while EMS calls are along the vertical axis. The pattern is the important aspect of a dot chart, rather than the individual dots. The horizontal scale of a dot chart should reflect the causation variable while the vertical scale reflects the resulting variable (that is, population causes or creates EMS calls).

> A **dot chart** reflects the pattern of one variable with another. The pattern is more important than the individual dots.

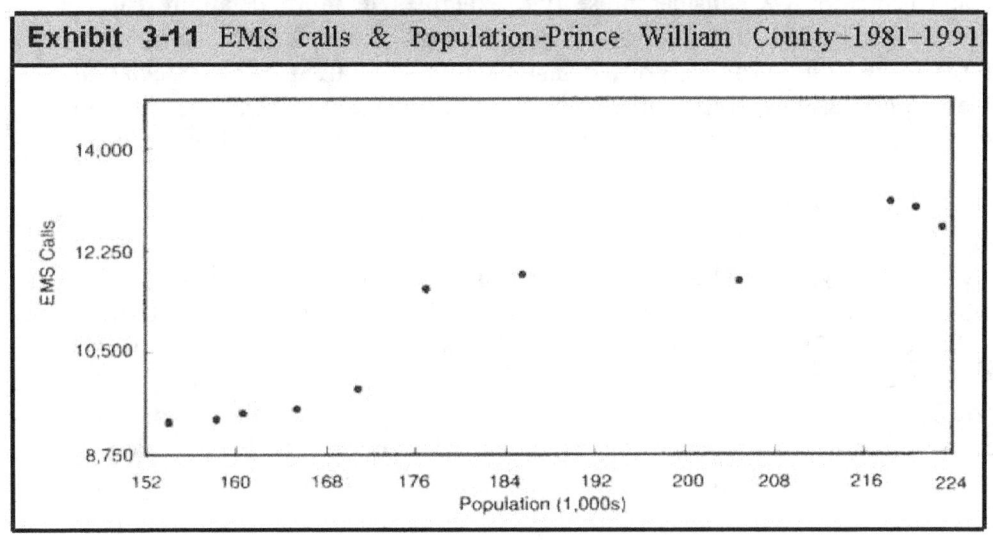

Exhibit 3-11 EMS calls & Population-Prince William County–1981–1991

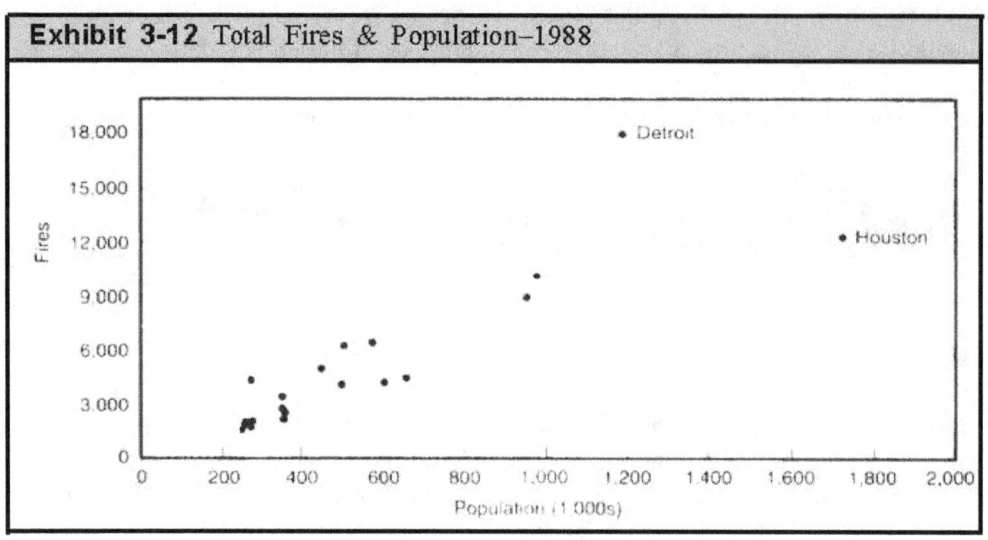

Exhibit 3-12 Total Fires & Population–1988

Another good application of scatter diagrams is to identify outliers in our data. In Chapter 2, we defined outliers as points that are isolated from the body of the data. Exhibit 3-12 shows diagram of population protected and total fires for 20 selected jurisdictions across the country. There is a general pattern showing the obvious fact that fires increase as population increases. However, there are two cities that do not follow this general pattern: Houston, Texas and Detroit. Michigan.

Detroit has more fires than expected based on its population, while Houston has fewer tires than expected based on its population. We make these conclusion because the dots for these two cities differ in location on the exhibit from the general pattern. The dot for Detroit is higher than we would expect from its polulation while Houston is lower than expected based on the general pattern.

In Chapter 5, we consider these two examples in more detail by calculating the correlation and regression line for each example. The correlation and regression line provide greater insight into the strength of association between population. EMS calls, and fires. Because a strong correlation exists, we can apply the regression line for future needs in fire departments.

Pictograms

Our final type of chart takes advantages of pictures to display data. Exhibit 3-13, for example, shows deaths per 1,000 tires for residential structure fires in 1988. Each contributing state has one of three distinquishing patterns reflecting low, average, and high rates. The overall rate of 8.3 injuries per 1,000 residential fires appears at the bottom of the chart. Low rates are in the .1 to 5.9 range (depicted) by light grey), avearge states are in the 6 to 11.9 range (medium grey). and high states have more than 12.0 deaths (dark grey).

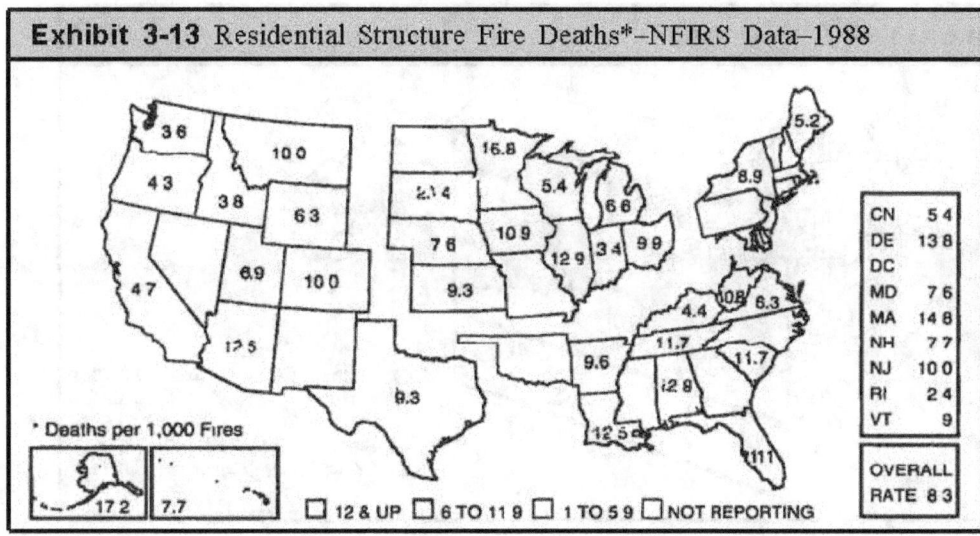

Exhibit 3-13 Residential Structure Fire Deaths*–NFIRS Data–1988

CN	5 4	
DE	13 8	
DC		
MD	7 6	
MA	14 8	
NH	7 7	
NJ	10 0	
RI	2 4	
VT	9	

OVERALL RATE 8 3

* Deaths per 1,000 Fires

☐ 12 & UP ☐ 6 TO 11 9 ☐ 1 TO 5 9 ☐ NOT REPORTING

The key is that presentation in this manner is more effective than any listing of the death rates. We can easily draw conclusions:

- States with high rates include Arizona, South Dakota, Minnesota, Illinois, Louisiana, Alabama, Delaware, and Massachusetts.
- Low rates predominate in the west (California, Oregon, Washington, and Idaho).
- A group of average states are in the midwest (Montana, Wyoming, Utah, Colorado, Texas, Nebraska, and Kansas).

Other charts for state and local data are easily imagined. At the state level, you may be collecting data from individual counties. A pictogram is a good way of depicting the county data by taking a state map showing county boundaries and developing an exhibit similar to Exhibit 3-13. Similarly, if you work for a local jurisdiction, such as a city or county, you may have data for individual fire districts. A jurisdiction map) with fire district boundaries may be an effective way of presenting the data.

As another example, Exhibit 3-14 on the following page shows areas of fire origins for residential structure fires in 1987 for the state of New York (excluding New York City). The percents appear in the tower right corner of the exhibit. The picture gives an effective way of highlighting where fires occurred within the residence.

A pictogram takes advantage of the background for the data you want to present. Data by geographical areas, such as counties, census tracts, or fire districts, can be presented on maps showing the boundaries of the areas. Similarly, data on structures, such as residences or manufacturing plants, can be presented on a schematic of the structure.

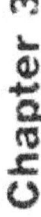

Chapter 3

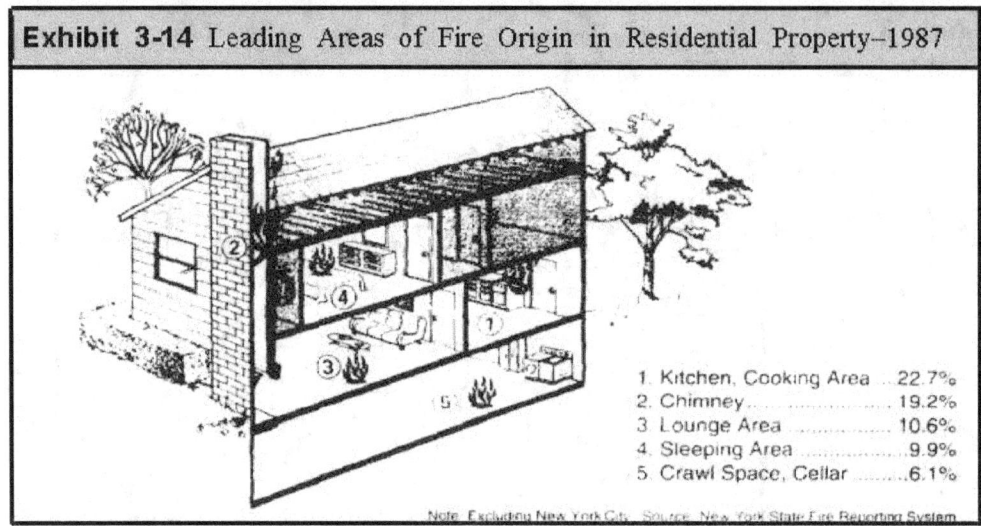

Exhibit 3-14 Leading Areas of Fire Origin in Residential Property–1987

1. Kitchen, Cooking Area ...22.7%
2. Chimney............................19.2%
3. Lounge Area 10.6%
4. Sleeping Area 9.9%
5. Crawl Space, Cellar 6.1%

Note: Excluding New York City. Source: New York State Fire Reporting System

Summary

In this chapter we presented six types of charts: bar charts, column charts, line charts, pie charts, dot charts, and pictograms. The primary purpose of using any chart is to indicate your conclusions more quickly and more clearly than is possible with tables and numbers. You may try several types of charts before you hit on the most appropriate. Be sure not to make your final chart too complicated. The message is what is important, so the chart form should not interfere.

As a quick reference guide on chart selection, we suggest the following:

- Use a **bar chart** when you have categorical data and your objective is to show how the items in a category rank. Most fire data is in categories, such as ignition factor, complex, type of ignition, form of ignition. These are reflected by the NFPA 901 codes.
- Use a **column** or **line chart** when you have data with a natural order, such as hours, months, or age groups. The chart will reflect the general pattern and indicate points of special interest, such as spikes, holes, gaps, and outtiers.
- A **pie chart** is beneficial when the objective is to show how the components relate to the whole. However, we suggest caution with pie charts. Keep the number of components to six or less and avoid forming several pie charts for comparison purposes.
- A **dot chart** depicts the relationship between two variables. Generally these variables are continuous rather than categorical. Population, travel times, and ages are examples of continuous variables. The pattern between the two variables is the important aspect for a dot chart.
- A **pictogram** is a pictorial representation of the data. Breakdowns by geographic areas are effectively shown by a pictogram.

1. The following are monthly data for 1990 from Los Angeles County, California showing the number of structure fires and vehicle fires for 1990. Develop charts for these figures and draw conclusions about the differences in the two distributions.

Month	Structure Fires	Vehicle Fires
January	115	159
February	122	151
March	124	173
April	83	136
May	111	142
June	132	183
July	130	176
August	103	171
September	85	187
October	108	163
November	138	128
December	154	145
Total	1,405	1,914

2. Through the NFIRS system, data are also collected nationwide on civilian injuries and deaths. The following data show age groups for male and female civilians injured in fires during 1989. Develop a chart comparing these distributions and draw conclusions about the differences.

Age	Males	Females
Less than 5 years	138	118
6 to 10 years old	64	35
11 to 15 years old	55	35
16 to 20 years old	83	55
21 to 25 years old	142	87
26 to 30 years old	152	123
31 to 35 years old	147	82
36 to 40 years old	125	54
41 to 45 years old	98	41
46 to 50 years old	71	43
51 to 55 years old	53	16
56 to 60 years old	55	39
61 to 65 years old	56	49
Over 65 years old	121	138
Total	1,360	915

(continued from question 2)

      a. Would a pie chart be appropriate for comparing the age groups?
      b. would you want to separate the data by sex?

3. The following data give the ages of firefighters injured or killed during 1989

Age of Firefighters	Number Injured
Under 20 years old	32
21 to 2 5 years old	172
26 to 30 years old	366
31 to 35 years old	426
36 to 40 years old	402
41 to 45 years old	253
46 to 50 years old	130
51 to 55 years old	60
Over 55 years old	35
Total	1,876

a.   Develop a chart for these data and make conclusions from the chart.

b. Now combine the figures into four age groups: up to 30 years of age, 31 to 40 years old, 41 to 50 years old, and than than 50 years old. Graph these groups with a pie chart.

4. Make charts and comparisons from the following data from four cities on types of fires found during 1990.

Type of Fire	Chicago	Dallas	Detroit	Phoenix
Structure	7,509	2,623	8,144	1,681
Vehicle	9,107	2,663	3,788	1,858
Trees, Brush, and Grass	1,154	2,046	435	2,155
Refuse	14,973	2,611	5,277	3,133
Total	32,743	9,943	17,644	8,827

5. The following figures and percentages are for fires in Los Angeles County for 1990 and Seattle, Washington 1990. An advantage of percentages is that they allow for quicker comparisons between distributions. Compare the percentages between Seattle and Los Angeles by day of week and provide some reasons for the similarities and differences.

Los Angeles County, 1990: Day of Week

Day	Number	Percent
Sunday	855	14.7
Monday	929	16.0
Tuesday	817	14.1
Wednesday	816	14.1
Thursday	778	13.4
Friday	767	13.2
Saturday	838	14.5
Total	5,800	100.0

Seattle, Washington, 1990: Day of Week

Day	Number	Percent
Sunday	358	14.6
Monday	338	13.7
Tuesday	359	14.6
Wednesday	368	15.0
Thursday	324	13.2
Friday	350	14.2
Saturday	362	14.7
Total	2,459	100.0

6. In 1990, the Chicago Fire Department responded to 9,107 vehicle fires. The following shows the amount of incident time in 5-minute increments for the fires. Develop a histogram for these incident times.

Minutes	# of Incidents	Percent
5-10	234	2.5
10-15	1,238	13.6
15-20	2,589	28.4
20-25	2,736	30.0
25-30	1,173	12.9
30-35	654	7.2
35-40	176	1.9
40-45	105	1.2
> 45	202	2.2

Chapter 3: Problems

7. During 1991, the Memphis Fire Department responded to 500 structure fires in which cooking was the cause of the fire. 'The following is a randomly selected sample of 30 fires from these 500 fires. The data show the incident times (from time of dispatch to time in service) and the dollar losses for these sampled fires.

Incident Time	Dollar Loss	Incident Time	Dollar Loss
6	50	23	500
9	700	25	1300
11	450	26	1500
11	500	27	135
12	1500	28	1800
13	250	29	700
14	500	32	700
15	600	32	1400
16	250	36	4500
17	100	40	4000
17	400	43	1800
18	300	44	15000
19	1000	45	300
19	1500	60	3000
21	300	78	1800

a. Develop a scatter plot for the 30 incidents.
b. Identify one obvious outlier in the data.
c. Redo the scatter plot without the outlier.
d in general, how arc incident times related to dollar losses?

Chapter 4
BASIC STATISTICS

Types of Variables

For purposes of analysis, fire department variables can be divided into two types: categorical variables and continuous variables. Categorical variables are defined as variables that are classified into groups or categories. For example, fires can be classified into structure fires, vehicle fires, refuse fires, explosions, etc. Categorical variables are sometimes called qualitative variables since we are not measuring quantities, but instead are classifying data into groups. Other examples of categorical variables are day of week, hour of day, month, zip code, fixed property use, ignition factor, method of alarm, area of fire origin, equipment involved in ignition, type of material ignited, construction type, extent of flame damage, and method of extinguishment. Most categorical variables are defined by the NFPA 901 codes, which form the foundation for reporting fire data into the NFIRS system, as described in Chapter 1.

Continuous variables always take on numerical values that reflect some type of measurement. Travel time to fires is an example of a continuous variable. We measure the travel time from the time the first unit is dispatched to a fire to the time the first unit arrives at the scene. It should be noted that the first unit to arrive may not be the same as the first unit dispatched. Other examples of continuous variables are on-scene time at fires, incident time at fires (travel time plus on-scene time), and dollar losses of fires. On-scene time is usually measured as the time from arrival of the first unit to the fire until the time back in service of the last unit. The on-scene time can range from practically no time, if the fire has extinguished itself, to several hours for a major fire.

We should note that if we say the on-scene time is 125 minutes, for example, it is highly unlikely that we are precisely correct. The time could be 124.7 minutes, 125.2 minutes or some other value close to 125 minutes. Most communication centers record individual times in hours and minutes, and we are certainly willing to accept on-scene time to the nearest minute for our analysis.

> **Categorical** variables are variables that are divided into group or categories. Days of the week, months and types of fires are examples of categorical variables.
> **Continuous** variables take an numerical values. Travel times, on-scene times, and dollar losses of fires are examples of continuous variables.

Chapter 4

In this chapter we will explore will how to summarize categorical and continuous variables. We will discuss what we mean by an average for a set of data and what we mean by variation of our data about an average.

It should be noted that we have been careful with the words **variable** and **data.** A variable is a characteristic that varies or changes. Days of the week vary from Sunday through Saturday; months vary from January through December; and types of fires vary according to NFPA 901 codes such as structure tires, vehicle fires, etc. Whenever we make observations on a variable, we develop data to be analyzed. Each time we complete a fire report, we create data by listing the day of week, hour of day, month, type of situation found, and values for all the other variables in the fire report. The data can then be summarized in a variety of ways, such as histograms and charts. In this chapter, we extend our ideas about summarizing data by introducing averages, standard deviations, box plots, and other techniques. In addition, Chapters 5 and 6 are devoted to analysis of categorical variables through two-way and three-way tables.

Averages: Mean, Median, and Mode

We saw in Chapter 2 and 3 that graphs provide good pictures of an entire set of observations, but we need more descriptive statistics for analysis purposes. In this section, we will develop averages for categorical and continuous variables. An **average** is a single number summarizing a set of data. It represents in a very general manner a "typical" data point. As seen in this section, there is more than one way to calculate an average. In fact, we discuss three different average: mean, median, and mode. Each has advantages and disadvantages, as explained as we go along. Always keep in mind that the aim is to develop a single number, called an average, that best describes the data.

Probably the most commonly known average is the **mean**, or **simple**

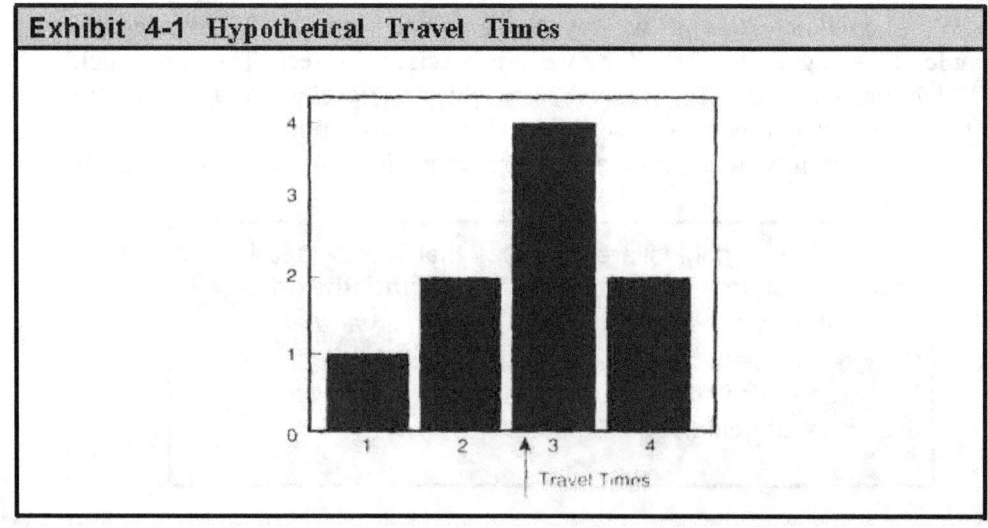

| Exhibit 4-1 Hypothetical Travel Times |

average, which is calculated by summing all the data values and dividing by the number of observations. For example, suppose that the travel times to nine incidents are 3 minutes, 2 minutes, 4 minutes, 1 minute, 2 minutes, 3 minutes, 3 minutes, 4 minutes, and 3 minutes. Adding these travel times gives 25 minutes in total and dividing by 9 gives a mean travel time of 2.78 minutes. A histogram of these 9 travel times would look like Exhibit 4-1. Notice that the mean balances the histogram in a seesaw manner.

It should also be noted that a mean can only be calculated with continuous variables, not with categorical variables. We can therefore calculate, for example, mean travel time, mean on-scene time, and mean dollar loss for fires.

A histogram of continuous data balances at the mean a **mean** can be calculated for continuous variables, but not for categorical variables.

Another type of average is the **median,** which is defined as the middle value (or the 50 percent point) in a group of numbers. For example, we have nine data values for our travel times. If we arrange these in order, they would look as follows: 1, 2, 2, 3, 3, 3, 3, 4, 4. The median is the fifth or middle value, which is 3 minutes in this example. There are four data values to the left of this number, and four values to the right. Another way of saying this is that 50 percent of the data is to the left of the median and 50 percent is to the right. For this reason, the median is also called the 50th percentile.

If we have an even number of data values, then there are two middle values and the median is the mean of the two values. For example, suppose that the on-site times for 10 fire incidents are 12, 15, 17, 25, 27, 29, 32, 35, 37, and 42 minutes. The two middle values are 27 and 29, so the median is 28 minutes (27 + 29 divided by 2). Note that the median again splits the values with five of the data values less than the median and five greater than the median.

As with the mean, the median can only be calculated for continuous variables. WE can determine the median on-scene time at fires and the median dollar loss for fires. However, the "median type of fire" or the "median ignition factor" has no meaning since these are categorical variables.

Two other percentiles frequently calculated from an ordered list of numbers are the 25th and 75th percentiles. Twenty-five percent of the data points are below the 25th percentile and 75 percent are below the 75th percentile. These are called the lower and upper quartiles. The interquartile difference is the difference between these two quartiles (that is, the upper quartile minus the lower quartile). A small interquartile range indicates that the data arc clustered around the median, while a large range reflects a wider spread of the data. In the list of 10 fire incidents above, the 25th per-

Chapter 4

centile is 16 and the 75th percentile is 33.5. The interquartile range is 17.5 which means that half of the data is within a 17.5 point range.

> The **median** is the middle value of a group of numbers. The median is also called the 50th percentile since 50% of the data points are lower than this number and 50 percent are higher. The 25th percentile is called the **lower quartile** and the 75th percentile is called the **upper quartile.** the **interquartile difference** is the numerical difference between the upper and lower quartile.

A final type of average is called the **mode,** which is the most frequent data value in your frequency list. It is easily recognized as the peak in a histogram. In Exhibit 4-1, 3 minutes is the mode.

> The **mode** is the peak of a histogram.

The mode is the only one of our three averages that can be applied to both categorical and continuous data. With categorical variables, the mode is the group with the highest number of values.

Effects of Extreme Values

We now want to see how the mean changes with changes in the data. The mean for our nine travel times was 2.78 minutes. In Exhibit 4-2, we have pushed the two travel times of 4 minutes each to 6 minutes. The mean also moves to the right to maintain the balance. Similarly, the bottom part of Exhibit 4-2 shows a mean of almost 4.0 minutes after moving the two travel times to 9 minutes.

The median and mode do not change in Exhibit 4-2. Half the values are still below 3 minutes and half are above 3 minutes, and the peak remains at 3 minutes.

This discussion illustrates that the mean is sensitive to extreme values while the median and mode are not. The sensitivity is especially important with analysis of fire department data. For example, most fires take less than two hours to complete, but a few may take several hours. These extreme fires inflate the mean on-scene time considerably, but have little effect on the median and mode.

If you have this situation, you may want to analyze the fires in two separate groups of on-scene times. In statistical terminology, we use the term **bimodal distribution** to describe data that are really a combination of two types of variables with very different means. Splitting data into two groups is advisable in bimodal situations so that each group can

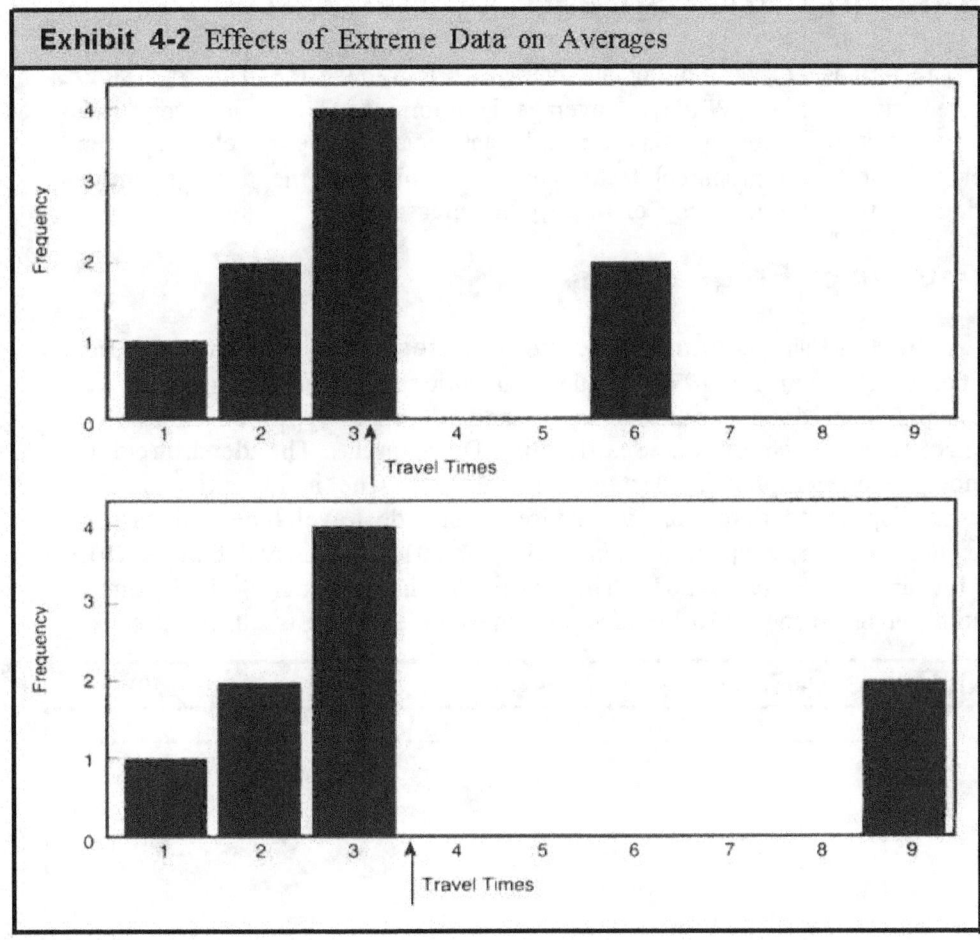

Exhibit 4-2 Effects of Extreme Data on Averages

be analyzed separately in a more meaningful manner.

Having introduced three different types of averages, the question usually arises as to which average is the best to use. Unfortunately, there is no single answer to this question. A good approach for selection of mean, median, or mode is to look at the distribution of the data, as indicated, for example, by a histogram. If the histogram shows that the data are clustered together with few extreme values, then either the mean or median is a good selection as an average value. On the other hand, we have just seen that extreme values inflate the mean. If extreme values are present in the histogram, then the median or mode may be the best average to represent the data. The mode is particularly useful if a large percentage of the data takes on the value of the mode.

Which is the vest average to use—mean, median, or mode? The answer is "It depends on the distribution of your data." The average to use is the number that is most representative of the data.

Chapter 4

Measuring the Spread of the Data

The purpose of developing an average is to reduce the data to a single representative number. While an average is informative, it is not very satisfactory by itself. Bearing this in mind, statisticians have developed several other measures and graphical techniques to supplement the average. In this section we will explore some of these other measures.

Cumulative Frequencies

Exhibit 4-3 shows **cumulative frequencies** for the incident times from Seattle, Washington for 1990. Incidents include fires, good intent calls, service calls. hazardous conditions calls, and all other types of calls which required a response by the Seattle Fire Department. The department responded to over 13,000 incidents during the year. The incident time is from time of dispatch to time hack in service. It include travel time and on-scene time. For **example**, suppose the alarm time is 2015, the arrival time is 2019, and the time in service is 2035. The travel time in this example is 4 minutes, the time at the scene is 16 minutes. and the incident time is 20 minutes.

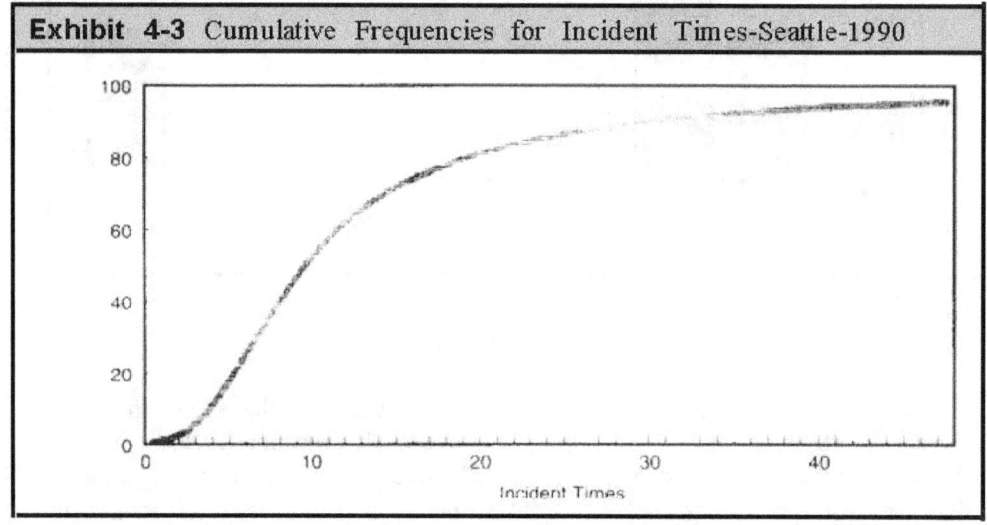

Exhibit 4-3 Cumulative Frequencies for Incident Times-Seattle-1990

You may remember that we introduced cumulative frequencies in Chapter 2. Cumulative frequencies tell you what percent of your data arc less than or equal to a given value. They are also useful for estimating the median of a distribution. In Exhibit 4-3, we can determine the median in the following manner. The left side of the exhibit shows percentages, and the median is the 50 percent mark. Starting at 50 percent we move across until we come to the curve and then move down to the incident time axis. This incident time is the median, which is about 10.2 minutes.

In a similar manner, the 25th percentile is about 6.5 minutes and the 75th percentile is about 17 minutes. This means that 25 percent of the incident times are 6.5 minutes or less and 75 percent are 17 minutes or less.

The interquartile range is 10.5 minutes since the upper quartile is 17 minutes and the lower quartile is 6.5 minutes.

Box Plot

A **box plot** is a graphical way of summarizing a continuous variable which takes advantage of percentiles, as shown in Exhibit 4-4.

In this diagram, the left side of the box is the 25th percentile and the right side is the 75th percentile. The vertical line inside the box is the median. Lines extended from each side of the box indicate the 10th and 90th percentiles (about 4.5 minutes and 30 minutes respectively).

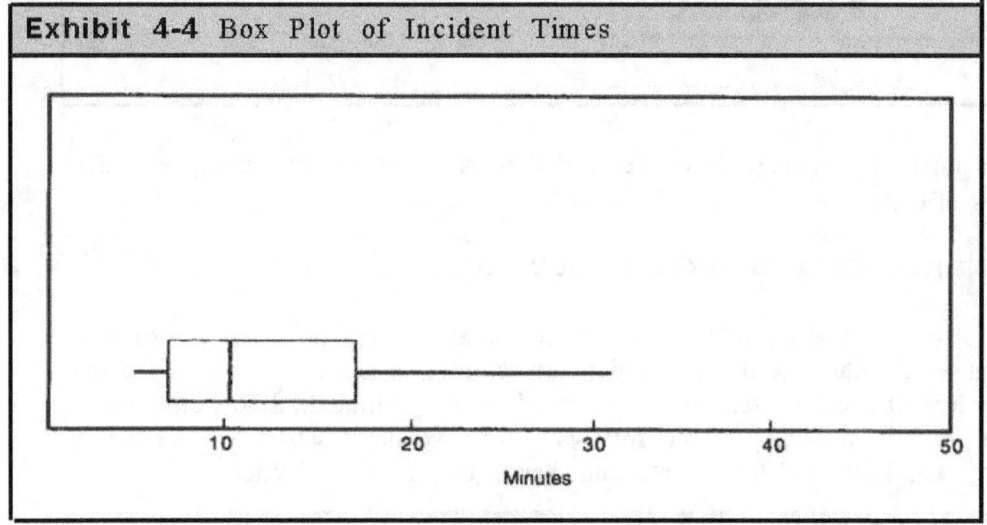

Exhibit 4-4 Box Plot of Incident Times

Minutes

A **box plot** is a graphical summary of a continuous variable. It displays key percentiles to indicate the median and overall spread of the data.

The long "whisker" extending from the right of the box says that 15 percent (from the 75th percentile to the 90th percentile) of the incident times are between 17 minutes and 30 minutes. Another 10 percent of the incident times are more than 30 minutes. Thus, the box plot indicates considerable variability in incident times.

One reason for the variability is the mix of incidents. During 1990, the Seattle, Washington Fire Department responded to 2,459 fires, 5,972 False Calls, 1,371 Service Calls, and 2,340 Good Intent Calls. These incidents have different average times because their on-scene activities differ. Exhibit 4-5 shows box plots for these four types of incidents. Good intent calls and false calls tend to have shorter incident times as indicated by lower medians and less spread in the data. Fire calls and service calls have larger medians and greater variability in their incident times.

Chapter 4

48

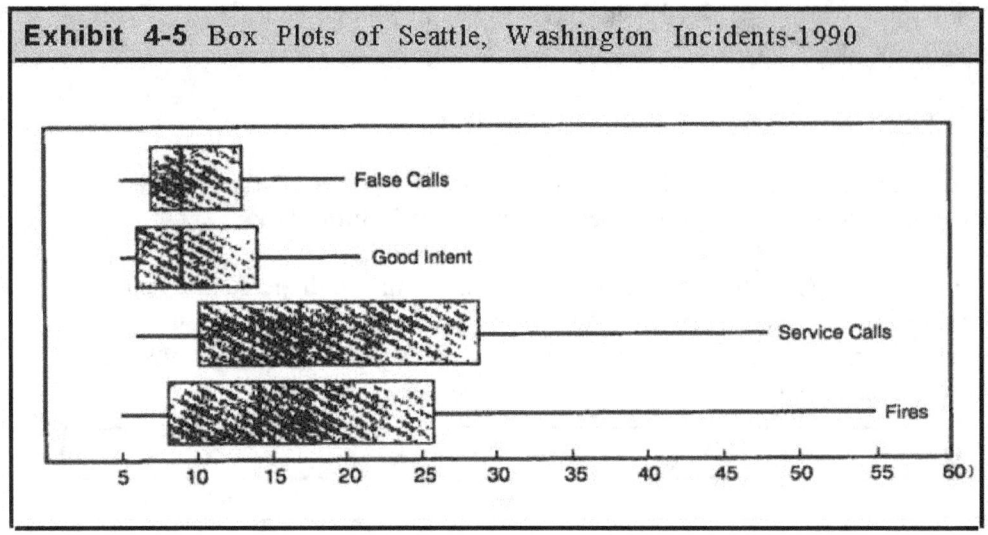

Exhibit 4-5 Box Plots of Seattle, Washington Incidents-1990

'The interquartile range is greater for fires and service calls than the other types of calls.

Variance and Standard Deviation

Another measure of the amount of spread in data is called the sample variance. To illustrate the calculation for sample variance, we go back to the nine hypothetical travel times discussed earlier which had a mean travel time of 2.78 minutes. In the following list, we have subtracted the mean from each individual travel time and then squared the difference.[2]

Exhibit 4-6 Calculation of Variance

Travel Time	Travel Time Mean (2.78)	Squared
1	-1.78	3.17
2	-.78	.61
2	-38	.61
3	.22	.05
3	22	.05
3	.22	.05
3	.22	.05
4	1.22	1.49
4	1.22	1.49
Total	0.00	7.57
Variance		.95

2. The square is the number multiplied by itself. For example. the square of 3 is 9 and the square of .6 is .36.

The middle column displays the amount of deviation from the mean for each point. The first deviation is -1.78 (1 minute minus 2.78 minutes), indicating that this travel time is 1.78 units from the mean and is to the left of the mean (since the sign is negative). Note that the sum of the middle column is zero; that is, the sum of the deviations from a mean adds to zero. In fact, an alternative definition for a mean is that it is the only number with this property; that is, it is the only number where the sum of the deviations is equal to zero.

In the right column, we square each deviation (that is, we multiply each deviation by itself). The sum of the squared deviations is 7.57 and the sample variance is obtained by dividing this sum by 8, which is one less than the total number of points. The variance is therefore 0.95. Since the variance is small compared to the mean, it indicates that the data points are close to the mean.

Finally, a statistic related to the variance is the sample standard deviation, which is the square root of the variance.' In our example, the standard deviation is .973, since this is the square root of .95. This means that the spread around the average is not very large (in this case less than 1.0 compared to a mean travel time of 2.8 minutes). The mean is therefore a good descriptor of the data in this example.

$$\text{Variance} = \frac{\sum (x_i - \overline{x})^2}{n-1}$$

$$\text{Standard Deviation} = \sqrt{\frac{\sum (x_i - \overline{x})^2}{n-1}}$$

For many continuous variables, the sample standard deviation has an interesting property - about 6.5 percent of the data values will be within one standard deviation of the mean. If the mean of a group of numbers is 50 and the standard deviation is 8, then about 65 percent of the data values will be between 42 and 58. Similarly, about 95 percent of the data values will be within two standard deviations.'

> The **sample variance** and **sample standard deviation** are measures of the spread of data. A small sample variance indicates that the data points are close to the mean. The sample standard deviation is the square root of the variance.

As an example, Exhibit 4-7 shows a histogram of travel times to incidents in Jersey City, New Jersey. The mean is 3.25 minutes and the standard deviation calculates to 1.5 minutes. The data in Exhibit 4-7 generally

3. The square root of a number means the number which multiplied by itself gives the original number. The square root of 4 IS 2; the square root of 25 IS .5, etc.

4. Most precisely, in statistical terms, the data values need to follow a normal distribution for this property to be true.

Chapter 4

Exhibit 4-7 Travel Times to Incidents-Jersey City, New Jersey

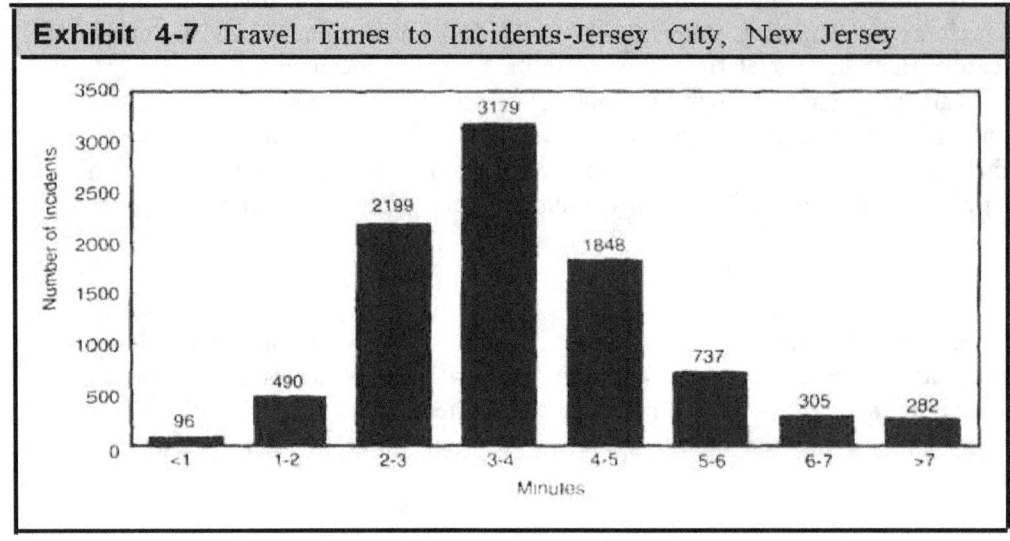

follows the pattern found with normal distributions. One standard deviation about the mean is from 1.75 minutes (3.25-1.5) to 47.5 minutes (3.25+1.5) About 68 percent of the travel times are within this one standard deviation range; about 95 percent of the travel times are within 2 standard deviations of the average.

Sample and Population Variances*

In the previous section, we used the terms sample variance and sample standard deviation. The application of the word sample is important here from a statistical viewpoint. We are assuming that the data we have are a sample of a larger set of data, such as the population of all incidents. When we say we have nine travel times, for example, we mean that we have a sample of nine travel times which we have selected to study. The travel times should be randomly selected so that they are representative of all travel times.

In algebraic terms, the sample variance is expressed as:

$$\text{Sample Variance} = s^2 = \frac{\sum (x_i - \bar{x})^2}{n-1}$$

The sample standard deviations(s) is the square root of the sample variance

If we are not working with a sample of data, but instead have the entire population, then the variance is calculated in a slightly different manner.

The **population variance**, usually designated as $\acute{o}^2$ is expressed on the following page.

$$\text{Population Variance} = \sigma^2 = \frac{\sum (x_i - u)^2}{N}$$

where N is the number of points in the population and a is the mean of the population. The **population standard deviation** (σ) is the square root of the population variance.

Note that the main difference between these two equations is that the sample variance is obtained by dividing by n-1 (the number of points in the sample), while the population variation is obtained by dividing by N (the total number of points).

With the calculation of a sample variance, we are really developing an estimate of the population variance. The reason for dividing by n-1 to obtain the sample variance is that statistical theory tells us that this gives a better estimate of the population variance than dividing by n. This difference may appear to he slight and insignificant, but it is important from a theoretical viewpoint.

If you are using a statistical package on a computer, it is important for you to know whether the output is the sample variance or the population variance. Unfortunately, many statistical packages will merely say "Variance" in the output without indicating whether it is the sample or population variance. One package may give the sample variance while the other may give the population variance. "The documentation to the package usually states which variance is calculated.

Indexes for Time Series*

Development of an index is another approach for understanding a frequency list of numbers. Indexes arc usually calculated for time series data, such as the distribution of fires by hour of day or by day of week. To develop an index, you divide each number in a series by the overall mean for the series. For example, Exhibit 4-8 shows the number of fires in 1990 in Chicago, Illinois. The table shows a total of 33,130 fires for the year which gives a mean of 2,760.8 fires per month. The column on the right is the index for each month. To develop the index for January, for example, we divide the number of fires in January by the mean:

$$\text{January Index} = \frac{2,406}{2,760.8} = .87$$

The indexes for the remaining months arc calculated in the same manner.

There are two interesting features of indexes. First, an index greater than I .0 means that the month is above average while an index below 1.0

Chapter 4

means the month is below average. This property is easy to see since we divided by the mean to get the index In Exhibit 4-8, we easily see that there are six months above average and six months below average. The second interesting feature of an index is that the index minus one gives the percentage that the month is above or below the mean. For example, July has an index of 1.25 and if we subtract 1.0 from this index, we can say that July is 25 percent above average. Similarly, January is I3 percent below average.

July	1.25 - 1 .00 = .25	25% above the mean
January	.87-1.00=-.13	13% below the mean

In summary, indexes are just another way of getting a feel for data. They are easy to calculate and easy to understand.

Exhibit 4-8 Index for Fires in Chicago–1990

Month	Total Fires	Index
January	2,406	.87
February	1,884	.68
March	2,522	.91
April	2,905	1.05
May	2,796	1.01
June	3,193	1.16
July	3,448	1.25
August	2,744	.99
September	3,118	1.13
October	2,689	.97
November	2,809	1.02
December	2,616	.95
Total	33,130	
Average	2,760.83	

Summary

In this chapter we have introduced several techniques for analyzing data. You should, at this point, understand how to construct a histogram and cumulative frequencies. You should also be able to estimate percentiles from cumulative frequencies and develop box plots.

The most useful statistics from a group of numbers are the mean, median, and mode. Be sure you understand how to calculate and interpret these numbers before you go further in this handbook. You should also understand how to calculate variance and standard deviations. These are particularly important in Chapter 7 on correlation and regression.

1.   The following data show travel time, on-scene time, and dollar loss for 27 hotel and motel fires (Fixed Property Use between 440 and 449) in Phoenix, Arizona during 1990.

Travel Time	On-Scene Time	Dollar Loss	Travel Time	On-Scene Time	Dollar Loss
4	26	$1,500	4	10	1,000
3	27	100	5	11	300
6	327	1,000	4	74	20,000
3	94	300	3	35	2,000
6	98	150	5	88	30,000
4	74	15,000	4	112	3,000
4	23	189	4	113	500
1	85	5,000	4	10	2,000
4	22	100	4	69	250
4	83	5,000	5	33	7,000
5	9	50	5	7	1,000
3	33	1,000	1	20	3,000
4	92	500	3	62	10,000
3	48	23,000			

a.   Calculate the mean, median, and interquartile range for each variable.

b.   Calculate the variance for travel time and dollar loss.

c.   How many dollar loss observations are within one standard deviation from the mean? How many within two standard deviations?

d.   Why is the variance for travel time so much less than for dollar loss?

e.   Comment on the effects of large dollar losses on the mean and median.

2.   Construct a histogram with the dollar loss data from the previous exhibit. From the histogram, construct a cumulative frequency distribution. Estimate the 10th, 25th, median, 75th, and 90th percentile from the cumulative frequency distribution.

54   **3.**   The following data show selected percentile for on-site time and dollar loss for fires in 519 one- and two-family dwellings (Fixed Property Use between 410 and 419) and 154 apartments, tenements, and flats (Fixed Property Use between 420 and 429) in Phoenix, Arizona during 1990.

On-Site Time

Percentiles

Type	10th	25th	50th	75th	90th
Dwelling	240	45.0	85.0	123.0	163.0
Apartments	20.5	32.0	71.5	107.0	143.0

Dollar Loss

Percentiles

Type	10th	25th	50th	75th	90th
Dwelling	1,500	3,000	10,000	25,000	48,000
Apartments	1,500	2,500	5,000	12,400	40,000

    a. Develop box plots for both variable on-site times, and dollar losses.

    b. Compare the interquartile ranges of dwellings versus apartments for the two variables.

    c. What are your conclusions from the box plots?

    d. Based on the box plot, would you expect a greater variance for dollar loss with dwellings or with apartments?

4.   In Chapter 2, we discussed skewness and showed that travel times were skewed to the right, i.e. , skewed toward high values. The Pearson coefficient of skewness has been proposed by some authors as a measure of the skewness of a distribution, Denoted by CS, the coefficient of skewness is defined as

$$CS= \frac{3(Mean - Median)}{Standard\ Deviation}$$

    a. Calculate the CS for the on-scene times and dollar losses in problem 1 above.

    b. When will the CS be zero?

5. A trimmed mean is a compromise average between the mean and median which is useful when outliers are present. A trimmed mean is computed by first ordering the data values from smallest to largest, deleting a selected number of values from each end of the ordered list, and finally averaging the values not deleted.

    a. Using the dollar loss data from problem 1, calculate the trimmed mean by deleting the two largest and two smallest values.

6. The following travel time data arc from Monroe Country, New York.

Cumulative Travel Times: Monroe County-1990

Cumulative Travel Time	Cumulative Frequency	Frequency	Percent
Less than 1 minute	157	157	7.5
1 to 2 minutes	287	444	21.2
2 to 3 minutes	324	768	36.7
3 to 4 minutes	322	1,090	52.2
4 to 5 minutes	271	1,361	65.1
5 to 6 minutes	227	1,588	76.0
6 to 7 minutes	176	1,764	84.4
7 to 8 minutes	120	1,884	90.1
8 to 9 minutes	99	1,983	94.9
9 to 10 minutes	45	2,028	97.0
10 to 11 minutes	25	2,053	98.2
11 to 12 minutes	16	2,069	99.0
12 to 13 minutes	11	2,080	99.5
13 to 16 minutes	10	2,090	100.0
Total	2,090		

    a. Plot the cumulative distribution percentages. Then develop a box plot with estimates from the cumulative distribution.

    b. By looking at your chart, what percent of travel times are under 8 minutes?

    c. Estimate the median using your chart.

Chapter 4: Problems

Chapter 5
ANALYSIS OF TABLES

Introduction

As we discussed in Chapter 1, the NFPA 901 Codes form the basis for coding fire and injury reports. They cover all the main variables associated with fires, including type of fire, fixed property use, ignition factor, type of complex, area of fire origin, type of material ignited, extent of flame damage, and many others.

For each variable, the 901 Codes provide specific categories, as exemplified by the categories for ignition factors shown in Exhibit 5-1:

Exhibit 5-1 Ignition Factor	
Code	Category
1	Incendiary
2	Suspicious
3	Misuse of Heat of Ignition
4	Misuse of Material Ignited
5	Mechanical Failure, Malfunction
6	Design, Construction, Installation Deficiency
7	Operational Deficiency
8	Natural Condition
9	Other Ignition Factors

Note that there is a numeric code for each category. However, from an analysis viewpoint, the code numbers have no statistical meaning. It is not reasonable to say that "Misuse of Heat of Ignition" is one more than "Suspicious" or to observe that "Design, Construction, Installation Deficiency" is twice "Misuse of Heat of Ignition." It is wrong in just the same way to perform statistical analysis with code numbers by calculating means and standard deviations. We cannot, for example, calculate the "Average Ignition Factor. " Thus, the code numbers tell us nothing statistically, but instead serve as a convenience when entering data into a computer. That is, it is much easier to enter a single number instead of entering the category name. There is also a considerable savings in the amount of computer storage required for the data when we can use a single number rather than storing a long name.

We make use of these codes by generating tables showing the number of observations (e.g., fires) for each category. We might find, for example, that the incendiary category accounts for 17 percent of the ignition, factors of fires, suspicious accounts for 23 percent, and so on.

In this chapter we will provide basic techniques for analyzing tables developed from categorical data. The first part of the chapter describes the development and interpretation of percentages for categorical data. We then develop a statistical test, called the chi-squared test, for determining whether the percentage distribution from a table differs significantly from a distribution of hypothetical or population percentages.

To summarize the terminology in this chapter, **variable** refers to a characteristic of fires and fire injuries. Each variable has several **categories** with a numeric code for each category. 'The codes have no statistical meaning, but assist in getting the data into a computer for analysis. Tables derived from the categories serve as the basis for calculating percentages and performing chi-squared tests of significance.

> A **categorical variable** divides a variable into a group of categories. Each category has a code assigned to it for convenient entry and storage in a computer. **Tables** can be developed showing the number of observations for each category. From the tables, we can calculate percentages and perform chi-squared tests of significance.

Describing Categorical Data

To summarize a categorical variable, we usually report the number of observations in each catepory- and its percentage of the total. For example, consider Exhibit 5-2 for types of situations found in the fires of' Seattle, Washington during 1989. These percentages arc simple to calculate and easy to understand: 36.3 percent of the fires are structure fires, 25.9 percent are vehicle fires, and so on. As described in Chapter 2, the mode is the category with the largest number of data values. In this example, the mode is structure fires, totaling 1,195 fires.

Exhibit 5-2 Type of Situations Found-Seattle Fires-1989			
Code	Type of Fire	Number	Percent
11	Structure Fires	1,195	38.3
12	Outside of Structure Fires	164	5.0
13	Vehicle Fires	850	25.9
14	Trees, Brush, Grass Fires	525	16.0
15	Refuse Fires	512	15.6
XX	Other Fires	42	1.3
	Total	3,288	100.0

Note that the percentages will not always add up to exactly 100 percent. This is because of rounding the individual calculations. The rounding may cause the total to be slightly more or slightly less than 100.0 percent. In this exhibit, the total is actually 100.1 percent, but we show 100.0 for convenience and consistency.

By way of comparison, Exhibit 5-3 shows the nationwide picture of types of situations found for fires. From a national perspective, structure fires accounted for 32.8 percent of the total, followed by vehicle fires at 23.9 percent, and trees, brush, and grass fires at 2 1.1 percent.

Exhibit 5-3 Type of Situations Found-Nationwide Fires-1989

Code	Type of Fire	Number	Percent
11	Structure Fires	302,708	32.8
12	Outside of Structure Fires	26,436	2.9
13	Vehicle Fires	220,861	23.9
14	Trees, Brush, Grass Fires	195,121	21.1
15	Refuse Fires	162,835	17.6
x x	Other Fires	15,922	1.7
	Total	923,883	100.0

A reasonable question to ask is whether the distribution of fires in Seattle differs from the national picture. We notice some differences by comparing percentages. For example, 36.3 percent of the Seattle fires are structure tires, compared to 32.8 percent nationwide. Similarly, 25.9 percent of the Seattle fires are vehicle fires, compared to 23.9 percent nationwide. We therefore suspect that the distribution of fires in Seattle deviates from the national picture, but a statistical test can be made to test this difference more precisely. In the next section we will provide such a test.

The Chi-squared Test

The chi-squared test is a statistical test to determine whether a sample set of observations has a different distribution from a hypothetical or population set of observations. The test is usually stated in more precise statistical language by defining an hypothesis to be tested. For our purposes, the null hypothesis, H_0, is that the percentage distribution from the sample distribution does not differ significantly from the national percentages. The **alternative hypothesis, H_1,** is that there is a significant difference between the two distributions.

To illustrate these ideas, we will deviate from our usual practice of showing examples with fire data. Instead, we will consider a simple experiment where we throw a die over and over again. The resulting data values are the number of dots showing after each throw. The number of dots

Chapter 5

varies between 1 and 6; that is, we have six possible outcomes. If u-e throw a "fair" die a large number of times, we would expect that about one-sixth of them would result in I dot showing, one-sixth in 2 dots showing, etc. The chi-squared test allows us to determine with some assurance whether we have a fair die.

> The **chi-squared test** determines whether a given distribution differs significantly from an hypothesized or population distribution. The null hypothisis is that no difference exists. while the alternative hypothesis is that there is a difference.

Suppose that we throw the die 90 times and obtain the results in Exhibit 5-4.

Exhibit 5-4 Results of Die Throws

Dots Visible	Number	Percent
One	16	17.8
Two	17	18.9
Three	12	13.3
Four	14	15.6
Five	17	18.9
Six	14	15.6
Total	90	100.0

If the die is a fair die, we should expect to have one dot visible exactly 15 times (one-sixth of the total), two dots visible exactly 15 times, and so on. Our actual results differ from these expected results as shown in Exhibit 5-5

Exhibit 5-5 Actual and Expected Results

Dots Visible	Actual Number	Expected Number
One	16	15
Two	17	15
Three	12	15
Four	14	15
Five	17	15
Six	14	15
Total	90	90

To summarize, we have tossed a die 90 times and obtained the results shown in Exhibit 5-4. Out null hypothesis is that the die is fair, which means that we expect the outcomes to be equally likely at one-sixth of the throws resulting in each possible outcome. The actual results are not the same as the expected either because of variations inherent in throwing a die only 90 times or because the die is not a fair die. The chi-squared test will determine whether the actual results differ significantly from the expected results.

To perform the chi-squared test, we do the following steps:

1. Calculate the **expected number** for each category by multiplying the expected or population percentages by the total sample size. This calculation has already been performed as shown in Exhibit S-S with the "Expected Number" column.
2. For each category, subtract the expected number from the actual number, and then square the result.
3. Divide the results from Step 2 by the expected number.
4. Sum the results from Step 3. This is the **calculated chi-squared statistic.** The larger this number, the more likely there is a significant difference between the observed and expected values. However, the chi-squared statistic also depends on the number of categories, which must be taken into account in the following steps.
5. Find the **degrees of freedom,** which is defined as the number of categories minus 1. In our die example, there are 5 degrees of freedom.
6. Obtain the **critical chi-squared value** from Appendix A by selecting the entry associated with the degrees of freedom. Now compare the computed chi-squared statistic from Step 4 to the critical value.

If the computed chi-squared statistic is **greater** than the value in the table, then we **reject** the null hypothesis. Otherwise, we accept the null hypothesis. To reject the null hypothesis means that the two distributions differ significantly. To accept the null hypothesis is to say that the two distributions are essentially the same with differences due to sampling or random variations.

> The **expected value** is the number we would expect if the null hypothesis were true. The chi-squared value is calculated by first subtracting the expected number from the observed number, squaring the result, and then dividing by the expected number. The results from these calculations are then summed to obtain the **calculated chi-squared value.**

Exhibit 5-6 on the following page summarizes these steps for the die example. The "Diff." column shows the difference between the expected and actual numbers. The "Squared Diff." is the square of the difference obtained by multiplying the number by itself. The right-most column is the

Chapter 5

Exhibit 5-6 Actual and Expected Results Die Tossing Experiment					
Dots Visible	Actual Number	Expected Number	Diff.	Squared Diff.	Divided by Exp.
One	16	15	1	1	.067
Two	17	15	2	4	.267
Three	12	15	-3	9	.600
Four	14	15	-1	1	.067
Five	17	15	2	4	.267
Six	14	15	-1	1	.067
Total	90	90			1.34

Chi-squared Value	1.34
Degrees of Freedom = 6 - 1	5
Critical Chi-squared Value	11.07

squared difference divided by the expected number; for example, the first figure is .067 obtained from 1 divided by 15.

The chi-squared value is 1.335, which is the sum of the values in the last column. In summary, the chi-squared value is given by:

$$\text{Chi-squared Value} = \sum \frac{(\text{observed} - \text{expected})^2}{\text{expected}}$$

From Appendix A, the critical chi-squared value for 5 degrees of freedom is 11.07. Since the calculated chi-squared value of 1.335 is less than this value, we accept the null hypothesis. That is, the results from the ninety throws do not provide evidence that the die is unfair.

Degrees of freedom have been defined as the number of categories minus one. The reason for this definition is as follows. Each category may be considered as contributing one piece of information or one degree of freedom to the chi-squared statistic. The exception is the last category which is not considered to he free because the total sample size is a fixed number. Consequently, the last category can he determined from the total sample size and the numbers in the other categories. Thus, the values in all categories except one can take on any values. We can see this same situation by looking at the calculated percentage. We know that the percentages must total to 100 percent. If the percentages are known except for one category, we can immediately calculate the percentage for the one remaining category?. If we have four categories, the first three percentages are free to vary, but the percentage for the last category is automatically determined. If the first three percentages are 25 percent, 30 percent and 35 percent, then the last category must be 10 percent so that the total sums to 100 percent.

> **Degrees of freedom** for a single list of items are equal to the number of items minus one. The term derives from the fact that all the items can vary except for one item (since the total is fixed).

We can now return to our question on whether the distribution of fires in Seattle differs from the nationwide distribution of fires. We noted differences in some categories; for example, Exhibit 5-2 shows that structure fires account for 36.3 percent of the fires in Seattle compared to 32.8 percent nationwide. Similarly, vehicle fires account for 25.9 percent of the fires in Seattle compared to 23.9 percent nationwide.

However, these are individual comparisons. The chi-squared test allows us to test all categories simultaneously. Our null hypothesis is that "The percentage distribution of fires in Seattle does not differ significanty from the nationwide picture" against the alternative hypothesis that "The percentage distribution of fires in Seattle differs from the nationwide picture." If the calculated chi-squared value is larger than the appropriate value in Appendix A, we will reject the null hypothesis: otherwise, we cannot reject the hypothesis.

Exhibit 5-7 shows the calculations using the information in Exhibits 5-2 and 5-3.

Exhibit 5-7 Actual and Expected Results-Seattle Fires-1989

Type of Fire	Actual Number	Expected Number	Diff.	Squared Diff.	Divided by Exp.
Structure	1,195	1,078.4	116.6	13,595.6	12.6
Outside	164	95.4	68.6	4,706.0	49.3
Vehicle	850	785.8	64.2	4,121.6	5.2
Grass	525	693.8	-168.8	28,493.4	41.1
Refuse	512	578.7	-66.7	4,448.9	7.7
Other	42	55.9	-13.9	193.2	3.5
Total	3,288	3,288.0			119.4

Chi-squared Value	119.4
Degrees of Freedom=6 - 1	5
Critical Chi-squared Value	11.07

Chapter 5

The "Actual Number" column comes directly from Exhibit 5-2. To obtain the expected number, we apply the percentages from Exhibit 5-3 to the 3,288 Seattle fires. For example, 32.8 percent of the nationwide fires were structure fires, which means wc **expect** 32.8 of the 3,288 fires in Seattle to be structure fires. This calculation gives 1 ,078.4 fires (32.8 percent times 3,288 fires).

The "Diff." column gives the difference between the actual and expected numbers and the next column is the squared difference (the difference multiplied by itself). The last column is the squared difference divided by the expected value. The calculated chi-squared value is the sum of the column, which is 119.4.

In this example, we have six categories of fires, which means we have five degrees of freedom. From Appendix A, the critical ch-squared value is 11.07. Since our calculated chi-squared value of 119.4 is greater than the critical value, we reject the null hypothesis. Our conclusion is that the distribution of fires in Seattle differs from the nationwide picture.

Other information is available from Exhibit 5-7 with regard to these differences. For example, the difference between the actual and expected number of structure fires is 116.6. Squaring this difference and dividing by the expected number gives 12.1, as shown in the last column. Even though there is a fairly large difference in this category, its contribution to the chi-squared value is not large. We expect large categories to have greater numerical variation than small categories, and for this reason, the calculation of the chi-squared value uses counts rather than percentages. On the other hand, the main contributors to the large chi-squared value are outside fires (49.3) and grass fires (41. 1). With Grass Fires, the difference between the actual and expected numbers is large (ignoring the minus sign) relative to the base values and the relatively low volume of fires creates the large contribution to the chi-squared value.

As another example of the ch-squared test, Exhibit 5-8 shows the number and percent of calls by month handled by a group of departments in Florida. This exhibit shows a flat distribution with roughly the same number of incidents each month. An exactly even distribution of calls would assign 8.33 percent of the calls to each month (100.0 percent divided by 12).

We can use the chi-squared procedure to test whether the observed distribution deviates significantly from an expected distribution having the same percent of incidents each month. Exhibit 5-9 shows the calculations for this test. The null hypothesis is that the actual distribution of incidents per month is the same as an expected distribution reflecting the same percent of incidents each month.

---

5 These numbers include all types of calls for which the fire department responded fire calls. rescue calls. hazardous conditions service calls. good Intent calls. false calls. etc

Exhibit 5-8 Fire calls for Selected Cities in Florida-1989

Month	Number	Percent
January	1,193	8.9
February	1,082	8.1
March	1,197	9.0
April	1,109	8.3
May	1,096	8.2
June	1,139	8.5
July	1,060	7.9
August	1,065	8.0
September	1,070	8.0
October	1,123	8.4
November	1,105	8.3
December	1,126	8.4
Total	13,365	100.0

Exhibit 5-9 Actual and Expected Results-Fire Call in Florida-1989

Month	Actual Number	Expected Number	Diff.	Squared Diff.	Divided by Exp.
January	1,193	1,113.7	79.3	6,288.5	5.65
February	1,082	1,113.7	-31.7	1,004.9	0.90
March	1,197	1,113.7	83.3	6,938.9	6.23
April	1,109	1,113.7	-4.7	22.1	0.02
May	1,096	1,113.7	-17.7	313.3	0.28
June	1,139	1,113.7	25.3	640.1	0.57
July	1,060	1,113.7	-53.7	2,883.7	2.59
August	1,065	1.113.7	-48.7	2,371.7	2.13
September	1,070	1,113.7	-43.7	1,909.7	1.71
October	1,123	1,113.7	9.3	86.5	0.08
November	1,105	1,113.7	-8.7	75.7	0.07
December	1,126	1,113.7	12.3	151.3	0.14
Total	13,365	13,365.0			20.37

Chi-squared Value	20.4
Degrees of Freedom = 12 - 1	11
Critical Chi-squared Value	24.7

In this case, the calculated chi-squared value (20.4) is less than the critical chi-squared value (24.7). We therefore accept the null hypothesis.

Chapter 5

That is, we conclude that the distribution of incidents by month does not differ significantly from an equal distribution.

This conclusion makes sense because of the weather conditions in Florida. The temperahire does not vary as greatly between winter and summer as in other states. For this reason, there may not he as much seasonal variation in fire department activity.

Technical Notes About the Chi-squared Test*

The chi-squared statistic is based on the **chi-squared distribution,** which is a well-known distribution to theoretical statisticians. If the null hypothesis, Ho, is true and if the sample size is sufficiently large, then the sampling distribution of the calculated chi-squared value is approximately the chi-squared distribution. There is a different chi-squared distribution for each degree of freedom. The primary attribute of the chi-squared statistic is that it tends to be large when the observed percentages are different from the hypothesized values. We reject the null hypothesis, which states they are all equal, when the calculated ch-squared value is "larger than reasonable."

The table in Appendix A is based on a 5 percent level of significance. To understand what we mean by this level of significance, we need to consider the die throwing experiment again. If we were to throw the die another 90 times, our results would probably be different. In fact, in some instances, the results would lead to a large chi-squared value, leading to a rejection of the null hypothesis of a fair die **even though the die was, in fact, fair.** This is called a Type 1 error. At first glance, this would appear to be a curious result, but is an accepted fact of life in statistical situations where random fluctuations play an important role.

When we state that a test is conducted at a 5 percent level of significance, we are saying that we expect to make an incorrect decision 5 percent of the time. That is, random fluctuations in the sampling procedures will result in an incorrect decision 5 percent of the time.

The choice of a 5 percent level of significance is common among statisticians in social science testing. It is not, however, always necessary) to select the 5 percent level. We can use a Type 1 error of 1 percent level, 10 percent level, or even 20 percent. The selection depends on the amount of risk we are willing to take in drawing an incorrect conclusion. At the I percent level, the critical ch-squared valuer, will be larger since we are saying that we will risk a wrong decision only 1 percent of the time. The critical values for the I percent level are presented in Exhibit 5-10.

Exhibit 5-10 Critical Chi-squared Values-1 Percent Significance Level	
Degrees of Freedom	Critical Chi-squared Values
1	6.6
2	9.2
3	11.3
4	13.3
5	15.1
6	16.8
7	18.5
8	20.1
9	21.7
10	23.2
11	24.7
12	26.2

A review of these figures shows that a larger calculated chi-squared value is needed in order to reject the null hypothesis. A 20 percent level provides smaller critical chi-squared values, but selecting such a high level is usually foolish since it means we risk a wrong decision 20 percent of the time.

As a final note, the chi-squared test should not be used with small samples. It is an excellent test when the sample size is large, but is not valid when the sample size is small. As the sample size increases, statistical theory says that the computed chi-squared statistic follows more and more closely to a chi-squared distribution when the null hypothesis is true. The best rule of thumb is to avoid using the chi-squared test when fewer than 5 cases are expected in any category. In some situations, it may be advisable to lump small categories together into an "Other" category prior to the chi-squared test.

Two-way Tables

In this section, we will extend our ideas to two categorical variables. By studying the two variables together rather than separately, we can measure the **statistical association** between the two variables. By association, we mean that knowing the value of one variable gives us information about the other variable. In some instances, there may he no association whatsoever, and we will be able to recognize this situation. Another result may be that the association exists, but is weak. Finally, the association between two variables may be quite strong, and we will measure the strength of this association.

Chapter 5

Exhibit 5-11 will serve as the starting point for to to introduce the concepts in this chapter. NFIRS data for 1989 on civilian injuries (not including deaths) was the source for developing the figures in this table. Each record in the database contains information on the location of the person at the time of the injury and the nature of the injury.

There are four categories for location. The first indicates that the person was intimately involved with the fire ignition. A common example is a person who is) burned as a result of accidentally spilling grease on a kitchen stove. This category also includes injuries from ignition of clothing on a person and from ignition of bedding or furniture on which a person is sitting or lying. The next category covers situations whcre the injured person is in the same room or space of the fire. but was not directly involved in the ignition. An example would be smoke inhalation by someone in the kitchen, but not directly at the stove when the grease fire occurred. The last two categories cover when the injured person is either on the same floor as the fire origin or on a different floor of the building.

Exhibit 5-11 Civilian Injuries-1989-Location and Nature of Injury

	Nature of Injury				
Location	Burns Only	Smoke Only	Burns & Smoke	Other	Row Totals
Fire Casualty intimately Involved with Ignition	238	34	116	6	394
Fire Casualty in the Room or Space of Fire Origin	130	100	100	10	340
Fire Casualty on Same Floor as Fire Origin	36	205	110	39	390
Fire Casualty in Same Building as Fire Origin	33	190	79	42	344
Column Totals	437	529	405	97	1,468

The nature of the injury is also divided into four categories (1) burns: only, (2) smoke/asphyxia only, (3) burns and smoke/asphyxia, and (4) other. The first three categories are self-explainatory. "Other" category includes injuries from shock, cuts. dislocations, fraction, and complaints of pain.

Exhibit 5-11 shows that a total of 1,468 injured persons. The top left number means there were 238 persons who were intimately involved in the fire and suffered injuries of burns only. This number is, in fact, the mode of the two-way classification (although the mode does not always have to appear as the first number in the table).

With the exception of identifying the mode, the numbers in the table do not relay much information. In the next section, we will calculate various percentages from this table, which will provide more insight. 'Then we will calculate a chi-squared value to measure the strength of the relationship between the two variables.

Percentages for Two-way Tables

There arc three different ways to calculate percentages for a two-way table of counts. Each way highlights a different feature of the table. More importantly, each provides a different interpretation of the data and leads to different conclusions about the relationship between the two variables. The three ways of calculating percentages are:

- Joint percentages
- Row percentages
- Column percentages

You select the type of percentage you want depending on what you are trying to show from the data. Joint percentages allow you to compare the entries in the table directly with each other. Ron percentages concentrate on each row of the table with percentages along that row summing to 100.0. In a similar manner, column percentages fix on each column of the table with percentages down the column summing to 100.0.

> Three types of percentages are possible with two-way tables: joint percentages, row percentages, and column percentages. **Joint percentages** are formed by dividing each number in the table by the grand total. **Row percentages** are obtained by dividing each number in a row by the row total, and **column percentages** are obtained by dividing each number in a column by the column total.

Joint Percentages

To develop joint percentages, we divide each entry in the table of counts by the overall total. Exhibit 5-12 shows the calculation for the counts from Exhibit .5- 11. The top left entry is simply:

$$\frac{238}{1468} = 16.2 \text{ percent}$$

This tells us that 16.2 percent of the total persons injured were intimately involved in the fire's ignition and suffer-cd burn injuries. The sum of all the entries in the table is 100.0 percent.

With the calculation of joint percentages, we have increased our knowledge because we can now make more logical comparisons. The table tells us, for example, that 14.0 percent of the persons were on the same floor as the fire origin and had smoke injuries. In a similar manner, only 2.2 percent of the persons were in the same building as the fire origin and had burn injuries.

Exhibit 5-12 Civilian Injuries-1989-Joint Percentages

	Nature of Injury				
Location	Burns Only	Smoke Only	Burns & Smoke	Other	Row Totals
Fire Casualty Intimately Involved with Ignition	16.2	2.3	7.9	.4	26.8
Fire Casualty in the Room or Space of Fire Origin	8.9	6.8	6.8	.7	23.2
Fire Casualty on Same Floor as Fire Origin	2.5	14.0	7.5	2.7	26.6
Fire Casualty in Same Building as Fire Origin	2.2	12.9	5.4	2.9	23.4
Column Totals	29.8	36.0	27.6	6.6	100.0

Exhibit 5-12 also provides important information from the row and column totals. For example, from the first row we find that 26.8 percent of the persons injured were intimately involved in the ignition. 'This percent can be derived in two ways. One way is to add the four percentages across the row (7.9 + 16.2 + 2.3 + 0.4 = 26.8). The other way is to divide the row total of 394 (from Exhibit 5-1I) by 1,468 to give 26.8 percent. Exhibit 5-12 shows a rather equal distribution across the four location categorics:

- 26.8 percent intimately involved in ignition
- 23.2 percent in the room or space of fire origin
- 26.6 percent on the same floor as the tit-c origin
- 23.4 percent in the same building as fire origin

In a similar manner, we have column percentages which provide information. For example, 29.8 percent of the persons injured suffered from burns only, 36.0 percent from smoke only, 27.6 percent with burns and smoke, and 6.6 percent with other injuries.

While Exhibit 5-12 provides mot-c insight into these two variables, it does not directly address other questions. For example, we cannot immediately compare burn injuries with smoke injuries for persons in the same room or space of fire origin. Similarly, we cannot compare persons located on the same floor with persons located in the same building for other injuries. These comparisons require calculation of row and column percents, as described in the following sections.

Row Percentages

To convert a table of counts into row percentages, we divide each entry in the table of counts by its row total. The top left entry is calculated by:

$$\frac{238}{394} = 60.4 \text{ percent}$$

This tells us that 60.4 percent of the total persons who were intimately involved in the fire ignition suffered burn injuries.

A table of row percentages allows for comparisons among the categories represented by the rows. The total for each row is 100.0 percent, and this figure appears on the right of the table as a reminder that we have developed row percentages.

As indicated, 60.4 percent suffered burn injuries when they were intimately involved with the fire's ignition. A total of 29.4 percent had burn and smoke injuries, 8.6 percent had smoke injuries and only I.5 percent had other injuries. These percentages account for all the injuries of persons intimately involved in the fire's ignition.

Exhibit 5-13 Civilian Injuries-1989-Row Percentages

	Nature of Injury				
Location	Burns Only	Smoke Only	Burns & Smoke	Other	Total
Fire Casualty Intimately Involved with Ignition	60.4	8.6	29.4	1.5	100.0
Fire Casualty in the Room or Space of Fire Origin	38.2	29.4	29.4	2.9	100.0
Fire Casualty on Same Floor as Fire Origin	9.2	52.6	28.2	10.0	100.0
Fire Casualty in Same Building as Fire Origin	9.6	55.2	23.0	12.2	100.0
Overall	29.8	36.0	27.6	6.6	100.0

Selecting the third row, which is for persons injured on the same floor as the fire's origin, a different picture emerges. Burns and smoke injuries account for 52.6 percent of the total, followed by 28.2 percent for burn injuries, and about 10.0 percent for the other two injury categories. Once again, these percentage total to 100.0 percent to account for all persons injured while on the same floor as the tire's origin.

It should he noted that we have repeated the overall percentages along

Chapter 5

72 the last row from the table of joint percentages. We can then make comparisons of the categories against the overall figures. For burn injuries, the overall percentage was 29.8 percent, and Exhibit 5- I3 shows that the first two location categories are above this figure while the last two location categories are below it.

Column Percentages

To convert a table of counts into column percentages, we divide each entry) by the total for its column. The top left entry would he calculated as:

$$\frac{238}{437} = 54.5 \text{ percent}$$

This tells us that 54.5 percent of the persons **who received burns** were intimately involved in the fire's ignition.

Exhibit 5-14 Civilian Injuries-1989-Column Percentages					
	Nature of Injury				
Location	Burns Only	Smoke Only	Burns & Smoke	Other	Total
Fire Casualty Intimately Involved with Ignition	54.5	6.4	28.6	6.2	26.8
Fire Casualty in the Room or Space of Fire Origin	29.7	18.9	24.7	10.3	23.2
Fire Casualty on Same Floor as Fire Origin	8.2	38.8	27.2	40.2	26.6
Fire Casualty in Same Building as Fire Origin	7.6	35.9	19.5	43.3	23.4
Overall	100.0	100.0	100.0	100.0	100.0

With a table of column percentages, we analyze a particular type of injury across the four locations. With burn injuries, we see that 54.5 percent occurred when the person was intimately involved with the fire's ignition. A total of 29.7 percent occurred when the person was in the room or space of the fire' origin, and less than 10 percent occurred in the other two location categories.

With the "Other" injury category, the picture changes. A total of 43.3 percent occurred when the person was in the same building as the fire's origin, followed closely by 40.2 percent for location on the same floor as the fire's origin. The first two location categories account for 10.3 percent and 6.2 percent, respectively.

Selecting a Percentage Table

The choice of a percentage table depends on what you are trying to conclude from the data. Joint probability tables are beneficial when you want to emphasize the interrelationship between the two variables in the table. Exhibit 5-12 shows that the combination of burns and intimate involvement in the fire's ignition account for 16.2 percent of the total. We can compare this figure to other combinations in the table.

The row percentage table provides a way of emphasizing the type of injury for each location. When the person was in the Same room or space of the fire's origin, Exhibit 5-13 shows 38.2 percent of the injuries were burns, 29.4 percent were smoke, 29.4 percent were burns and smoke, and 2.9 percent were other injuries. These are useful results by themselves, and can be compared to distributions in other rows.

The column percentage table emphasizes the location for each type of injury. For burns only, Exhibit 5-14 shows that 54.5 percent were intimately involved in the fire's ignition, 29.7 percent were in the same room or space of the fire's origin, 8.2 percent on the same floor, and 7.6 percent in the same building.

Testing for Independence in Two-way Tables

In this section we will develop a &i-squared test for testing whether the two variables in a two-way table are independent of each other. As with our prior discussion, we will provide a step-by-step procedure for calculating a chi-squared value in a two-way table. We would like to note at this point, however, that virtually all statstical packages automatically calculate the chi-squared value for you. As you may have concluded with the examples from Seattle and Florida, manual calculation is arduous and time consuming. In practice, it is not advisable to figure out chi-squared values with pencil and paper. However, we go through an example in detail in this section so that you understand what a statistical package is doing when it calculates a chi-squared value. You should appreciate the time saved and the inherent accuracy of these packages in your applications.

Before getting to chi-squared calculations, however, we need to know what we mean by independence. We say that two variables are **independent** if knowledge about one variable does not help us in predicting the outcome of the other variable. In the table of location versus type of injury, we should certainly suspect that the two variables are not independent. If we know, for example, that the person was intimately involved in the tire's ignition, then we can predict that the person probably had burn injuries. As we shall see later, the chi-squared test will confirm the dependence between our two variables.

Independence of two variables means that knowledge of one variable does not help in predicting the other variable. An equivalent definition is that two variables are independent if either their row or column percentages are equal.

As an example of two independent variables, consider the following table showing the relationship between sex and severity of injury in fires. The table shows a total of 1,841 persons who were either injured or killed as a result of a fire. Of this total, there were 1.092 males and 749 females. Of the 1,841 persons, there were 1,561 persons injured and 280 persons killed. This table includes the row percentages, which have rounded for purposes of illustration.

Exhibit 5-15 Sex and Severity of Injuries with Row Percentages-1989			
		Injury Severity	
Sex	Injured	Killed	Total
Male	927	165%	1,092
Row%	85%	15%	100%
Female	634	115	749
Row%	85%	15%	100%
Total	1,561	280	1,841
Row%	85%	15%	100%

This table shows identical row percentages for males and females. That is, 85 percent of the males and 85 percent of the females received injuries. Consequently, knowledge of the sex of a person does not improve our ability to predict injury severity. With either sex, the row percentages have the Same distribution of 85 percent for injured and 15 percent for killed.

This table illustrates an equivalent definition for independence: two variables are independent if either their row percentages or column percentages are the same. When the percentages agree, we have no predictive power.

We cannot always expect that the row or column percentages in a table will be so close that independence is as obvious as Exhibit 5-15. The chi-squared test with two-way tables provides a means to test whether two variables are independent. With a chi-squared test, we can determine in a statistical manner whether the variables are independent.

In order to perform the chi-squared test, we first need to develop **expected values** for the table. The expected values are the counts that would occur if the two variables were independent. After forming a table of expected values, we will be in a position to do the chi-squared test.

Table of Expected Values

We form an expected value table in the following manner. For each number in the original table of counts, we identify the associated row total and column total. The entry in the expected value table is formed by multiplying the row sum by the column sum and dividing by the grand total. We can state this calculation as follows:

$$\text{Expected Value} = \frac{\text{Row Sum x Column Sum}}{\text{Grand Total}}$$

As an example, look back at Exhibit 5-11, which is the table of counts for the location and nature of injury. From our prior analysis with row and column percentages, we strongly suspect an association between location and nature of injury. The chi-squared test allows us to confirm our suspicion statistically.

The top left entry shows 238 persons who were intimately involved in the fire's ignition and had burn injuries. The row sum associated with this value is 394, and the column sum is 437. The expected value is therefore calculated as:

$$\text{Expected Value} = \frac{394 \times 437}{1468} = 117.29$$

We continue to perform this type of calculate for every entry in the original table of counts. Exhibit 5-16 shows the resulting expected value table.

Exhibit 5-16 Civilian Injuries-1989-Table of expected Values

| | Nature of Injury | | | | |
Location	Burns Only	Smoke Only	Burns & Smoke	Other	Total
Fire Casualty Intimately Involved with Ignition	117.30	142.00	108.70	26.00	394.0
Fire Casualty in the Room or Space of Fire Origin	101.21	122.52	93.80	22.47	340.0
Fire Casualty on Same Floor as Fire Origin	116.10	140.54	107.60	25.77	390.0
Fire Casualty in Same Building as Fire Origin	102.40	123.96	94.90	22.73	344.0
Overall	437.00	529.00	405.00	97.00	1,468.0

Chapter 5

The table of expected values is what we would expect if location and nature of injury were completely independent. It should also be noted that the row totals and column totals arc exactly, the same as the original table of counts. That is, development of the expected value table preserves these totals.

With the table of expected values in place, we can proceed with describing the calculations for the chi-squared test.

Chi-squared Test for Two-way Tables

The chi-squared value is calculated along the same lines as we did for testing just a single categorical variable.

1. Develop the table of expected values, as shown in Exhibit 5-16. Each entry in the table is obtained by multiplying the row total times the column total and then dividing by the grand total.
2. For each table entry, subtract the cxpected value from the corresponding entry in the original table of counts, and then square the result. This difference measures the discrepancy between the actual counts and what we would expect under independence.
3. Divide the results from Step 2 by the expected value. This is an adjustment that allows for the fact that larger expected numbers are usually associated with larger deviations.
4. Sum the results from Step 3. This is the chi-sqnarcd statistic. The larger the chi-squared statistic, the more likely there is a independence between the two variables. However, the &-squared statistic also depends on the number of categories, which must he taken into account in the following steps.
5. Find the **degrees of freedom,** which is calculated for two-way tables by multiplying (number of rows minus one) times (number of columns minus one). For our example, we have four rows and four columns. The number of degrees of freedom is therefore ('+-I) x (4-1) = 9.
6. Compare the computed A-squared statistic from Step 4 to the value in the chi-squared table in Appendix A using the appropriate degrees of freedom. This table value is called the **critical chi-squared value.**

If the computed chi-squared statistic is **greater** than the value in the table, then we **reject** the null hypothesis. Otherwise, we cannot reject the null hypothesis.

It is important to keep in mind that the null hypothesis with a two-way table is that the two variables are independent. If we accept the null hypothesis, we are saying that knowing the value of one of the variables does not help in predicting the value of the other variable. In our example, the null hypothesis is that location is independent of the nature of the injury.

Exhibit 5-17 shows the chi-squared entries for our two-way table. These entries are the results after Step 3 above. The top left entry was calculated as follows: Exhibit 5-11 gave an actual count of 238 for this entry and Exhibit 5-16 gave an expected value of 117.30. Subtracting the expected value from the actual count gives 120.7 (238 minus 117.3) and squaring gives 14,568.49. Dividing this number by the expected value, 117.3 provides the entry for the chi-squared table of 124.20.

Exhibit 5-17 Civilian Injuries-1989-Table of Chi-squared Entries

Nature of Injury

Location	Burns Only	Smoke Only	Burns & Smoke	Other
Fire Casualty intimately Involved with Ignition	124.20	82.14	0.49	15.39
Fire Casualty in the Room or Space of Fire Origin	8.20	4.13	0.41	6.94
Fire Casualty on Same Floor as Fire Origin	55.26	29.61	0.05	6.75
Fire Casualty in Same Building as Fire Origin	47.04	35.13	2.66	16.41

Total Chi-squared Value = 434.81

To determine the total chi-squared value, we add the numbers from Exhibit 5-17, which gives 434.81.

We are now ready to test our hypothesis about independence of these two variables - location and nature of injury. From Appendix A, we see that the critical chi-squared value for 9 degrees of freedom is 16.92. Our calculated chi-squared value of 434.81 greatly exceeds the critical value.

We therefore reject the null hypothesis and conclude that there is a statistical association between location and nature of injury.

Chi-squared Calculation for 2 x 2 Tables*

One exception to the above steps occurs with 2 x 2 tables. For 2 x 2 tables, we must subtract 0.5 from positive differences between observed and expected counts and add 0.5 from negative differences. This exception is necessary in order for 2 x 2 tables to match the chi-squared distribution under the null hypothesis. In most tables, this adjustment makes little difference in the calculated chi-squared value. However, in some tables it is important and can change the outcome of the test. For this reason, the adjustment should always be made in 2 x 2 tables.

As an illustration of this exception, consider the following table showing sex by part of body injured for the 1989 data.

Exhibit 5-18 Sex and Part of Body Injured with Row Percentaged-1989			
	Part of Body Injured		
Sex	Internal	External	Total
Male	366	726	1,092
Row %	33.5	66.5	100.0
Female	338	411	749
Row %	45.1	54.9	100.0
Total	1,561	280	1,841
Row %	38.2	61.8	100.0

In this table, we have collapsed the injuries into internal (including respiratory and heart) and external (head, body, arm, leg, etc.) injuries. The row percentages indicate differences by sex. For males, 33.5 percent had internal injuries and 66.5 percent had external injuries while for females, the percentages are 45. 1 percent and 54.9 percent, respectively.

The chi-squared value from this table is 24.87. which has been adjusted according to the above rule on adding and subtracting 0.5. With a 2 x 2 table, there is always only one degree of freedom. From Appendix A, the critical value with 1 degree of freedom is 3.84. Since our calculated chi-squared value of 24.87 is greater than this critical value, we reject the null hypothesis.

We conclude that there is an association between sex and part of body injured.

PROBLEMS

1. Suppose that we roll a die 60 times with the following results:

Dots Visible	Number	Percent
One	13	21.7
Two	I2	20.0
Three	7	11.7
Four	12	20.0
Five	9	15.0
Six	7	11.7
Total	60	100.0

Use a chi-squared test to determine whether this die is a "fair" die at the 5 percent level.

2. In New Orleans, Louisiana, the number of fires per month for 1990 was as follows:

January	467
February	291
March	392
April	322
May	319
June	349
July	384
August	374
September	359
October	368
November	298
December	345
Total	4,268

a. With a chi-squared test at the 5 percent Ievel, determine whether there is an even distribution of fires in New Orleans.

b. Which months contribute significantly to deviations from an even distribution?

3. The data on the following pages are for 1990 fires for NFIRS Metropolitan areas and for the remainder of the United States. The NFIRS Metropolitan areas are a group of 25 of the largest cities contributing to NFIRS. The question is whether the distribution of types of fires in the metropolitan areas differs from the rest of the country.

Code	Type of Fire	Metro Areas	Rest of United States
11	Structure Fires	54,189	241,481
12	Outside of Structure Fires	4,378	23,986
13	Vehicle Fires	51,774	174,398
14	Trees, Brush and Grass Fires	28,440	172,217
15	Refuse Fires	59,767	111,291
	Other Fires	2,833	15,353
	Total	201,381	738,726

Source: NFIRS Tally Report 22, 1990.

a. Calculate percentages for each group and develop a tentative answer based only on the percentages.

b Use the chi-squared test to determine whether the two distributions are significantly different at the 5 percent level.

c. Which types of tires account for the main differences between the two groups?

4. The following data for days of the week are for 1990 fires for Denver, Colorado and for the entire United States. The percentages are for the United States. Using a chi-square test, determine whether the distribution by day of week for Denver differs from the entire country.

Day	Denver Fires	Fires in United States	Percent
Sunday	674	134,077	14.37
Monday	688	136,984	14.68
Tuesday	565	130,283	13.96
Wednesday	550	134,886	14.45
Thursday	576	128,729	13.79
Friday	588	129,522	13.88
Saturday	603	138,854	14.88
Total	4,244	933,335	100.0

5. The following data for days of the week are for 1990 fires for Denver, Colorado and for the other metropolitan cities. Using a chi-square test, determine whether the distribution by day of week for Denver differs from the other metropolitan cities.

Day	Denver Fires	Fires in Metro Cities	Percent
Sunday	674	29,382	15.18
Monday	688	28,961	14.96
Tuesday	565	27,200	14.05
Wednesday	550	27,305	14.10
Thursday	576	26,025	13.44
Friday	588	26,226	13.55
Saturday	603	28,505	14.72
Total	4,244	193,604	100.0

ADVANCED TABLE ANALYSIS

Introduction

In Chapter 5, we focused on statistical tools for analyzing the relationship between two variables. We now want to extend our ideas to situations where we need to understand the relationships among several categorical variables. For example, consider the problem of trying to determine factors that affect the spread of a fire beyond the initial room of origin. We might ask several key questions: If a building has fire detectors, is a fire less likely to spread to other rooms? To what extent does the type of material ignited affect the spread of a fire? If equipment is involved, is there greater likelihood or less likelihood of the fire spreading to other parts of the building?

Each of these variables-fire spread, presence of a fire detector, type of material, involvement of equipment-is a categorical variable. That is, each variable is divided into several categories as defined by the NFPA 901 codes. The box on fire incident reports labeled "Extent of Flame Damage" indicates fire spread according to seven categories defined by the NFPA 901 codes." Similarly, fire detector performance, type of material, and equipment involvement are also recorded on the fire incident report with NFPA 901 codes.

Loglinear analysis is a statistical approach for analyzing the relationships among several categorical variables. The complexities addressed by loglinear analysis are easily illustrated. We could, for example, develop several two-way tables from the four variables just described:

* Fire spread by fire detector use
* Fire spread by type of material ignited
* Fire spread by involvement of equipment
* Fire detector use by type of material ignited
* Fire detector use by involvement of equipment
* Type of material ignited by involvement of equipment

Each two-way table defines a relationship that may be important in understanding fire spread. We could proceed even further by defining three-way tables, such as fire spread by fire detector use by type of material ignited. This three-way relationship may also be important in explaining the spread of a fire. The point is that we have many potential relationships with these four variables. Loglinear analysis assists in identifying the important relationships.

---

6. The categories are (1) fire confined to object of origin (2) fire confined to part of room, (3) fire confined to room, (4) fire confined to fire-rated compartment of origin, (5) fire confined to floor of origin. (6) fire confined to structure of origin, and (7) fire extended beyond structure of origin.

In this chapter we will present an overview of loglinear analysis. We start by applying loglinear analysis to two-dimensional tables. This discussion serves as an introduction to the ideas behind this approach. Our primary example in this chapter is a table with four variable: extent of flame damage, fire detector performance, type of material ignited, and whether equipment was involved in the fire.

It should he mentioned that this chapter differs from previous chapters because it assumes some knowledge of statistics. You should be able to understand this chapter if you have had an introductory) course in statistics. In particular, an understanding of the standardized normal distribution is required. The calculations for loglinear analysis are also more difficult than the relatively simple calculations in previous chapters. However, all the statistical packages mentioned in Chapter 1 include loglinear analysis procedures which will automatically perform the calculations. Your job will be to interpret the results and select the most appropriate model.

> **Loglinear analysis** is a statistical technique for determining relationships among variables in multi-dimensional tables. From assumptions about interactions among table variables, loglinear analysis develops a model of the table and then tests expected results from the model against the actual table. You can develop several models with different assumptions and then select the model that most appropriately describes the relationships in the table.

Loglinear Analysis of 2 x 2 Tables*

Model of Independence

Exhibit 6-1 is a hypothetical table for two variables that arc mutually, independent. You can verify that the row percentages arc the same for both categories of variable A, which proves its independence with variable B. We have introduced notation in Exhibit 6-1 for identifying the individual cells and the sums along each row and column. For example, x_{11} denotes the first category of variable A and the first category of variable B. Similarly), x_{12} indicates the first category of variable A and the second category of variable B.

The expected value for a cell in the table can be expressed in the following manner:

$$m_{ij} = \frac{(x_{i+})(x_{+j})}{x_{++}} \qquad i = 1,2 \quad j = 1,2 \qquad (1)$$

where m_{ij} is the expected value for the cell at row i and column j. Equation (1) says that the expected value, m_{ij} is found by multiplying the total for row

$i(x_{i+})$ by the total for column $j(x_{+j})$ and then dividing by the grand total (x_{++}). This calculation is exactly the same as presented in Chapter 5 for determining expected values.

Exhibit 6-1 Hypothetical Tables for Two Independent Variables

	Variable B		
Variable A	Category B1	Category B2	Total
Category A$_1$	36 x_{11}	24 x_{12}	60 x_{1+}
Category A2	48 x_{21}	32 x_{22}	80 x_{2+}
Total	84 x_{+1}	56 x_{+2}	140 x_{++}

If we take the **natural** logarithm of both sides of (l), we obtain the following:

$$\log m_{ij} = \log x_{i+} + \log x_{+j} - \log x_{++} \tag{2}$$

To generalize our discussion, we will think in terms of a two-dimensional table with I rows and J columns. Equations (1) and (2) still hold as expressions of expected values and their logarithms.

The formal model of independence is expressed by rewriting (2) in the following manner.

$$\log m_{ij} = u + u_{A(i)} + u_{B(j)} \tag{3}$$

$$u = \frac{1}{IJ} \sum_i \sum_j \log m_{ij} \tag{4}$$

$$u_{A(i)} + u = \frac{1}{J} \sum_j \log m_{ij} \qquad i = 1,2,...I \tag{5}$$

$$u_{B(j)} + u = \frac{1}{I} \sum_i \log m_{ij} \qquad j = 1,2,...J \tag{6}$$

The term u is the overall mean of the logarithms of the numbers in the table. Similarly, $u_{A(i)} + u$ gives the mean of the logarithms of the expected counts for the J cells at level i of variable A and uB(j) + u is the mean of the logarithms of the expected counts for the I cells at level j of variable B.

Equations (3) through (6) represent the **loglinear model** of independence for two-dimensional tables. The name loglinear model derives from the fact that we have taken the logarithms of the expected values to form a linear combination. The model is valid for 2 x 2 tables as well as larger tables with I rows and J columns.

We can apply these equations to Exhibit 6-I to get a sense of how the loglinear model of independence operates. The results are shown in Exhibit 6-2. To obtain the value for u, apply Equation 4 to the cell entries:

$$u = \frac{1}{4} (\log(36) + \log(24) + \log(48) + \log(32)) = 3,525 \qquad (7)$$

Similarly, we obtain $u_{A(1)}$ and $u_{B(1)}$ by applying Equations (5) and (6):

$$u_{A(1)} = \frac{\log(36) + \log(24)}{2} \quad 3.525 = -.144 \qquad (8)$$

$$u_{B(1)} = \frac{\log(36) + \log(43)}{2} - 3.525 = .203$$

Exhibit 6-2 Values of Terms for Model of Independence

Term	Term Value
u	3.525
$u_{A(1)}$	-.144
$u_{A(2)}$	.144
$u_{B(1)}$	.203
$u_{B(2)}$	-.203

An important feature of the model is that the sum of uA(1), and uA(2) is equal to zero. Similarly, the sum of uB(1) and uB(2) is also equal to zero. The reason they sum to zero is that they are deviations from an overall average. We have had other discussions in this handbook where the sum of deviations from a mean is equal to zero. In general, we can state that:

$$\sum_i u_{A(i)} = \sum_j u_{B(j)} = 0 \qquad (9)$$

The values in Exhibit 6-2 can be employed with Equation (3) to obtain Exhibit 6-1. The logarithm of m21, the number of injured females, is shown on the following page.

$$\log m_{21} = u + u_{A(2)} + u_{B(1)}$$

$$\tag{10}$$

$$\log m_{21} = 3.525 + .144 + .203 = 3.872$$

The antilogarithm gives $m_{21} = 48$, which agrees with Exhibit 6-1.

Model of Dependence

As we saw in Chapter 5, we frequently encounter tables in which the two variables are not independent. As an example, we analyzed location and nature of injury in Chapter 5 quite extensively. Exhibit 6-3 reproduces the table so that we can illustrate the extension of our model to two dependent variables.

Exhibit 6-3 Civilian Injuries-1989					
	Nature of Injury				
Location	Burns Only	Smoke Only	Burns & Smoke	Other	Row Totals
Fire Casualty Intimately Involved with ignition	238	34	116	6	394
Fire Casualty in the Room or Space of Fire Origin	130	100	100	10	340
Fire Casualty on Same Floor as Fire Origin	36	205	110	39	390
Fire Casualty in Same Building as Fire Origin	33	190	79	42	344
Column Totals	437	529	405	97	1,468

The extension is accomplished by introducing **interaction terms. We** think of location and nature of injury as "interacting" and develop an extended model to take the interactions into account:

$$\log m_{ij} = u + u_{A(i)} + u_{B(j)} + u_{AB(ij)}$$

$$\tag{11}$$

where $u_{AB(ij)}$ represents the interactions between variables A and B. To calculate $u_{AB(ij)}$, we simply use equation (11) and solve for uAB(ij). Equations (4) thru (6) and Equation (9) still apply, and in addition, we: have:

$$\sum_i u_{AB(ij)} = \sum_j u_{AB(ij)} = 0$$

$$\tag{12}$$

Exhibit 6-4 shows the results for location and nature of injury. We have an overall mean, u, of 4.128 for the model. Since location has four cat-

Chapter 6

88

egories, we have four $u_{A(i)}$ terms. Similarly, we have four $u_{B(j)}$ terms for the four categories of injury. There are 16 terms for $u_{AB(ij)}$ to cover the 16 combinations of location and nature of injury. The terms $uA(i)$ and $u_{B(j)}$ are called the **main effects** of the model and the terms $u_{AB(ij)}$ are called the **two-way interaction effects.**

These terms operate in exactly the same way as with the independent model. That is, we can obtain an entry in Exhibit 6-3 by applying Equation 11. Exhibit 6-3 shows 205 persons with injuries of smoke only and location on the same floor as the fire's origin (x_{32}). To obtain this value from our model, we calculate as follows:

$$\log m_{32} = u + u_{A(3)} + u_{B(2)} + u_{AB(32)}$$
$$= 4.128 + -190 + .548 + .458 \qquad (13)$$
$$= 5.324$$

The antilogarithm gives the value of 205 persons.

Exhibit 6-4 Values of Terms for Dependent Model			
Location and Nature of Injury			
Term	Term Value	Term	Term Value
u	4.128		
$u_{A(1)}$	-.242	$u_{B(1)}$	.227
$u_{A(2)}$	-.033	$u_{B(2)}$	.548
$u_{A(3)}$	.190	$u_{B(3)}$	.479
$u_{A(4)}$	.085	$u_{B(4)}$	-1.254
$u_{AB(11)}$	1.359	$u_{AB(31)}$	-.961
$u_{AB(12)}$	-.907	$u_{AB(32)}$	.458
$u_{AB(13)}$	.388	$u_{AB(33)}$	-.096
$u_{AB(14)}$	-.840	$u_{AB(34)}$	.600
$u_{AB(21)}$	545	$u_{AB(41)}$	-.943
$u_{AB(22)}$	-.038	$u_{AB(42)}$	.487
$u_{AB(23)}$	.031	$u_{AB(43)}$	-.323
$u_{AB(24)}$	-.539	$u_{AB(44)}$	.779

With the model given by Equation (11) for a two-dimensional table, we will always obtain the exact value in the original table. For this reason, the loglinear model with all interaction terms is called a **saturated** model. Later in this chapter, we will develop **unsaturated models** by eliminating terms from the saturated model. The aim with unsaturated models is to obtain good estimates for a table with fewer terms than a saturated model.

While unsaturated models rarely produce the exact values in a table, they sometimes provide estimates close to the table values. The unsaturated model then has a key advantage of having identified the important interactions among the variables.

> For a two-dimensional table, the saturated loglinear model is expressed as log $m_{ij} = u + u_{A(i)} + u_{B(j)} + u_{AB(ij)}$, where u is the overall mean of the logarithms of table entries, $u_{A(i)}$ are the terms for the main effects of variable A, $u_{B(j)}$ are the terms for the main effects of variable B, and $u_{AB(ij)}$ are the interactions terms. If the two variables are independent, then the terms $u_{AB(ij)}$ equal zero and do not appear in the model.

As a final example in this section, Exhibit 6-5 shows a 2 x 2 table on civilian injuries relating sex of the injured person to external and internal injuries from fires. External injuries are usually burns to a part of the body while internal injuries are usually from smoke inhalation. Because the row percentages differ considerably, the two variables are clearly not independent. A saturated model is therefore appropriate for this table.

Exhibit 6-5 Civilian Injuries-Sex by Type of Injury-1989

	External	Internal	Total	
Male	362	331	693	$u = 5.716$
Row Percent	52.2	57.8	100.0	
				$u_{A(1)} = .131$
				$u_{A(2)} = -.131$
Female	211	336	547	$u_{B(1)} = -.094$
Row Percent	38.6	61.4	100.0	$u_{B(2)} = .094$
				$u_{AB(11)} = .138$
Total	573	667	1,240	$u_{AB(12)} = -.138$
	46.2	53.8	100.0	$u_{AB(21)} = -.138$
				$u_{AB(22)} = .138$

The right portion of the table shows the terms for the saturated model. Each entry in the table can be derived using Equation (11) with these terms.

One other feature of 2 x 2 tables is important for the remainder of this chapter. Referring back to Exhibit 6-1, suppose that we randomly select an individual from category B_1. The quantity x_{11}/x_{21} represents the **odds** of an individual appearing in category A_1 rather than A_2. Similarly, x_{12}/x_{22} gives the odds of an individual appeari in A_1 rather than A_2 **given** that the individual was from category B_2. In Exhibit 6-5, we have 573 persons with external injuries and the odds are 1.72 of a male rather than a female (362 divided by 211) having external injuries. We usually express the odds as 1.72: 1 of males to females. If internal injuries are involved, the odds are .98:1 (331 divided by 336) of males to females. Odds greater than 1.0 indicate that the first category is larger than the second category, while odds less than 1 .0 indicate the opposite.

The **odds ratio** is the ratio of these two odds. and can be written as:

$$\text{Odds Ratio} = \frac{x_{11}x_{22}}{x_{12}x_{21}} \tag{14}$$

If the two variables are independent, then the odds ratio will always be equal to 1.0. Conversely, an odds ratio of exactly 1.0 indicates complete independence of two variables. On the other hand, the odds ratio from Exhibit 6-5 is 1.74, which indicates that sex and type of injury are not independent.

Another way to derive the odds ratio is from probabilities. For example, Exhibit 6-6 shows probabilities calculated by dividing each entry in Exhibit 6-5 by the total of 1,240 persons. The odds ratio is obtained by $p_{11}p_{22}/p_{21}p_{12}$. For Exhibit 6-6, the calculation gives 1.74 (.292 x .271 / 170 x .267), which agrees with our previous calculation. The reason for showing this approach is that we may be given the probabilities, rather than the actual data, as our starting table. The calculations show that the results are the same regardless of the starting point.

Exhibit 6-6 Civilian Injuries-Sex by Type of Injury-Probabilities-1989			
	External	Internal	Total
Male	.292 p_{11}	.267 p_{12}	.559
Female	.170 p_{21}	.271 p_{22}	.441
Total	.462	.538	1.000

We will return to the idea of odds ratio later in this chapter when we present results from an example with four variables.

Three-way Tables and Standardized Values*

Suppose we have three variables labeled A, B, and C with I, J, and K categories, respectively. A three-way table with these categories would have $I \times J \times K$ cells. Each ceil would indicate the number of items with attributes Ai, Bj, and Ck. We consider the items to be a random sample taken from a population from which the talk is derived.

We let $w_{ijk} = \log_e x_{ijk}$, where X_{ijk} is the number of observations in ceil (i,j,k). The saturated model for a three-way table is given by:

$$w_{ijk} = u + u_{A(i)} + u_{B(j)} + u_{C(k)} + u_{AB(ij)} + u_{AC(ik)} + u_{BC(jk)} + u_{ABC(ijk)} \quad (15)$$

As with the previous discussion, u is the overall mean, and uA(i), uB(i), and uc(k) are the main effects. Similarly, uAB(ij), uAC(ik), and UBC(jk) represent the two-way interactions, and uABC(ijk) are the three-way interactions. For a $2 \times 2 \times 2$ table, we would have an overall mean, 4 terms for each of the two-way interactions, and 8 terms for the three-way interactions.

The sums of various terms in the model are zero:

$$\sum_i u_{A(i)} = \sum_j u_{B(j)} = \sum_k u_{C(k)} = \sum_i u_{AB(ij)} = \sum_j u_{AB(ij)} = \sum_i u_{AC(ik)} = \sum_k u_{AC(ik)} \quad (16)$$

$$= \sum_j u_{BC(jk)} = \sum_k u_{BC(jk)} = \sum_i u_{ABC(ijk)} = \sum_j u_{ABC(ijk)} = \sum_k u_{ABC(ijk)}$$

$$= 0$$

While Equations (15) and (16) appear complex, the calculations are relatively straightforward. For example, u is the average of the natural logarithms of the table entries:

$$(17)$$

$$u = \frac{1}{IJK} \sum_i \sum_j \sum_k w_{ijk}$$

To obtain the other values in the saturated model, we need the following definitions:

$$(18)$$

$$w_{i++} = \frac{1}{JK} \sum_j \sum_k w_{ijk}$$

$$(19)$$

$$w_{+j+} = \frac{1}{IK} \sum_i \sum_k w_{ijk}$$

$$w_{++k} = \frac{1}{IJ} \sum_i \sum_j w_{ijk} \tag{20}$$

Then we obtain:

$$u_{A(i)} = w_{i++} - u \qquad i = 1,2 \dots, I \tag{21}$$

$$u_{B(j)} = w_{+j+} - u \qquad j = 1,2 \dots, J \tag{22}$$

$$u_{C(k)} = w_{++k} - u \qquad k = 1,2 \dots, K \tag{23}$$

By extending the model, we develop terms for $u_{AB(ij)}$ and $u_{ABC(ijk)}$ as follows:

$$u_{AB(ij)} = w_{ij+} - w_{i++} - w_{+j+} + u \tag{24}$$

$$u_{ABC(ijk)} = w_{ijk} - w_{ij+} - w_{i+k} - w_{+jk} + w_{i++} + w_{+j+} + w_{++k} - u \tag{25}$$

As an example of this saturated model, the value for $u_{AB(11)}$ for a 2 x 2 x 2 table is given by:

$$u_{AB(11)} = w_{11+} - w_{1++} - w_{+1+} + w_{+++} \tag{26}$$

$$= \tfrac{1}{2}(w_{111} + w_{112}) - \tfrac{1}{4}(w_{111} + w_{112} + w_{121} + w_{122})$$
$$- \tfrac{1}{4}(w_{111} + w_{112} + w_{211} + w_{212})$$
$$- \tfrac{1}{8}(w_{111} + w_{112} + w_{121} + w_{122} + w_{211} + w_{212} + w_{221} + w_{222})$$

$$= \tfrac{1}{8}(w_{111}) + \tfrac{1}{8}(w_{112}) - \tfrac{1}{8}(w_{121}) - \tfrac{1}{8}(w_{122}) - \tfrac{1}{8}(w_{211})$$
$$- \tfrac{1}{8}(w_{212}) + \tfrac{1}{8}(w_{221}) + \tfrac{1}{8}(w_{222})$$

It should be noted that each number in the last equation is multiplied by either by or $\tfrac{1}{8}$ or $\tfrac{1}{8}$.

Upton (1978) shows that any parameter in a saturated model can be expressed in the following form:

$$\sum_i \sum_j \sum_k a_{ijk} w_{ijk} \tag{27}$$

where the aijk are suitably developed constants. In the example just given, the a_{ijk} are either $\tfrac{1}{8}$ or $-\tfrac{1}{8}$ depending on the particular cell.

Upton also discusses the fact that the variance of any is related to the original frequencies in the table, and can be approximated by:

$$\mathrm{Var}(w_{ijk}) = \frac{1}{x_{ijk}} \tag{28}$$

The key point is that we can approximate the variance of a parameter with the following relationship:

$$\text{Variance} = \sum_i \sum_j \sum_k \frac{(a_{ijk})^2}{x_{ijk}} \qquad (29)$$

In many applications, our aim will be to identify the most important terms in a model. The estimated variances will not always be the same since they depend on the number of categories. To compare the terms, we therefore need to standardize, which can be done as follows:

$$\text{Standardized Term} = \frac{u^*}{\sqrt{V(u^*)}} \qquad (30)$$

where u^* represents any of the terms in the model and $V(u^*)$ is the variance of the term according to Equation (27). Goodman (1971) shows that these standardized values follow an approximately normal distribution with an expected mean of 0 and a variance of 1.

The value of the above result is that it allows us to determine the importance of terms in a model. In the next section, we will show the standardized terms for a saturated model in a four-dimensional table and use the standardized terms to identify key interactions among the four variables.

Loglinear Analysis Applied to a Four-Dimensional Table: Fire Data in Chicago, Illinois

Saturated Model

With the background from the previous sections, we are in a position to analyze higher-dimensional tables. Our aim is to determine how several categorical variables relate to each other. That is, we want to examine the interactions among variables as a way of explaining the entire table.

As an example for this section, we use data from 3,548 residential fires that occurred in Chicago, Illinois in 1990.[7] We have selected four variables to study with the following definitions:

Variable	Description
A	Fire was confined to the room of origin, or was not
B	Detector performed, or did not (or was not present)
C	Fire started from fabric material, or did not
D	Equipment was involved in the fire, or was not

7. Residential fires classified as arsons are not included in this analysis.

The first variable (A) indicates whether the fire was confined to the room of origin or whether it extended to another room, another floor in the structure, or perhaps even another structure. Variable A is typically catted the extent of flame damage. The second variable (B) indicates whether a fire detector went off during the fire. The third variable (c) indicates whether the form of material ignited was a fabric (such as cotton, wool, fur, etc.) or another type of material (such as flammable liquids, plastics, wood, or paper). The last variable (D) indicates whether equipment was involved in the fire.8

All four variables are dichotomous variables; that is, they are divided into two categories as either present or not present. We intentional]!. developed dichotomous variables to simptify. the presentation of the loglinear analysis. Extent of flame damage actually has seven categories indicating whether the fire was confined to (1) object of origin. (1) parts of room or area of origin, (3) room of origin, (4) tire-rated compartment of origin, (5) floor or origin, (6) structure of origin, or (7) beyond structure of origin. For our analysis, the first four categories defined A1 (room of origin) and the last three catcgories define A2 (beyond room of origin). It should be mentioned that you can apply loglinear analysis to a table with all seven categories; however, each cell in the table should contain at least five fires in order to assure valid results from the analysis.

Exhibit 6-7 on the following page gives the breakdown of these 3,548 residential fires according to our four-way classification. The first category of each variable is coded as a one and the second category as a two. With detector performance, for example, a code of one indicates that a fire detector went off, while a code of two indicates a fire detector did not operate. The exhibit defines a 2 x 2 x 2 x 2 table with a total of 16 cells. Cell (l,l,l,l) indicates 37 fires where the fire remained in the room of origin, the detector went off, fabric was the type of material ignited, and equipment was also involved in the fire. Cell (2,2,2,1) contain 125 fires where the fire extended beyond the room of origin, a detector was not present, fabric material was not involved in the tire, and equipment was involved in the fire.

exhibit 6-8 shows the term values and standardized terms for the saturated model. For each u term, a subscript appears in the "Term" column indicating the variables. The term value and standardized value are shown in the nest two columns. They always correspond to the first category of a variable. The first item in the table is for the first category of variable A confined to room of origin). It shows a term value of $u_A = 555$ and a standardized value of 16.0. As shown in the previous section, term values sum to zero, and similarly, standardized values also sum to zero. We therefore know that the second category of variable A (fire extended beyond the room of origin) has a term value of -.555 and a standardized value of -16.0. The other figures in the exhibit operate in the same manner since all our variables are dichotomous.

---

8 A code of 98 in the Equipment Involved in Ignition indicates that no equipment was involved in the fire. All four variables have been rededined from the NFPA 901 codes to form dichotomous variables

Exhibit 6-7 Residential Fires in Chicago, Illinois – 1990

			Frequency	Extent of Flame Damage	Fire Detector	Definition Material	Equipment
Cell							
1	1	1	37	Room of Origin	Worked	Fabric	Equipment Involved
1	1	2	165	Room of Origin	Worked	Fabric	No Equipment Involved
1	2	1	218	Room of Origin	Worked	Other Material	Equipment Involved
1	2	2	181	Room of Origin	Worked	Other Material	No Equipment Involved
2	1	1	95	Room of Origin	Did Not Work	Fabric	Equipment Involved
2	1	2	504	Room of Origin	Did Not Work	Fabric	No Equipment Involved
2	2	1	487	Room of Origin	Did Not Work	Other Material	Equipment Involved
2	2	2	798	Room of Origin	Did Not Work	Other Material	No Equipment Involved
1	1	1	6	Beyond Room of Origin	Worked	Fabric	Equipment Involved
1	1	2	55	Beyond Room of Origin	Worked	Fabric	No Equipment Involved
2	1	1	35	Beyond Room of Origin	Worked	Other Material	Equipment involved
2	1	2	70	Beyond Room of Origin	Worked	Other Material	No Equipment Involved
2	1	1	46	Beyond Room of Origin	Did Not Work	Fabric	Equipment involved
2	1	2	262	Beyond Room of Origin	Did Not Work	Fabric	No Equipment Involved
2	2	1	125	Beyond Room of Origin	Did Not Work	Other Material	Equipment Involved
2	2	2	464	Beyond Room of Origin	Did Not Work	Other Material	No Equipment Involved

Exhibit 6-8 Standarized Values for Terms in Standard Model

Chicago Residential Fires-1990

Term	Term Value	Standardized Value	Term	Term Value	Standardized Value
A	555	16.0	BD	.067	1.9
B	-.688	-19.9	CD	-.295	-8.5
C	-.465	-13.4			
D	-.583	-16.8	ABC	.038	1.1
			ABD	.041	1.2
AB	.146	4.2	ACD	-.060	-1.7
AC	-.026	-0.6	BCD	-096	-2.6
AD	.151	4.4			
BC	-.009	-0.3	ABCD	.033	1.0

If we study the standardized values, we can identify the most important variables according to the saturated model. Since the standardized terms follow a normal distribution with mean zero and variance one, **a good rule of thumb is to look at standardized values with magnitudes greater than 2.0 (ignoring the sign).** Any standardized value with magnitude greater than 2.0 indicates a term particularly important to the model. The table shows that all four main effects (A, B, C, and D) are important. In addition, the following interaction effects, in order of absolute magnitude, arc important:

C D	Type of material and equipment involved
AD	Extent of flame damage and equipment involved
A B	Extent of flame damage and detector performance
BCD	Detector performance, type of material, and equipment involved

These results will be useful in the next section where we develop a log-linear model that contains only the interactions of importance.

Continuation of Chicago Example: Hierarchical Models*

By definition, saturated models include every possible interaction in a table. We can derive beneficial conclusions from a saturated model because it is an exact representation of the table. The question we explore in this section is whether we can eliminate terms from the saturated model and still obtain good estimates of the table values. With a reduced model, we will identify more clearly the important interactions in the table. Conversely, terms not included in a reduced model do not contribute to our understanding of the table.

Hierarchical models provide a structured approach to define models with fewer terms than saturated models. The rule with an hierarchical model is that any term included in the model automatically means that higher-level terms are also included. For example, if u_{AB} is included, then u_A and and u_B are automatically included. Of course, the overall mean, u, appears in all models. If u_{ABC} is in the model, then we include all two-way interactions and main effects: u_{AB}, u_{AC}, u_{BC}, u_A, u_B, and u_C. Hierarchical models are unsaturated models because they have fewer terms than saturated models.

With this approach, we are establishing a hierarchy of terms. We are saying that if an interaction term, such as u_{AB} is important, then the main effects, u_A and u_B are also important. Exhibit 6-9 shows the 9 possible hierarchical models for a table with three variables. The first model is the saturated model which we have already presented. The model A/BC consists of the main effects for all three variables and the two-way interaction term for variables B and C. The model AB/AC consists of the main effects for all three variables and the two interaction effects AB and AC, but excludes BC and ABC.

Exhibit 6-9 Possible Hierarchical Models for Tables with Three Variables		
Model	Model Terms	Degrees of Freedom for 2 x 2 x 2 Model
ABC	u, u_A, u_B, u_C, u_{AB}, u_{AC}, u_{BC}, u_{ABC} (Saturated Model)	0
AB/AC/BC	u, u_A, u_B, u_C, u_{AB}, u_{AC}, u_{BC}	1
AB/AC	u, u_A, u_B, u_C, u_{AB}, u_{AC}	2
AB/BC	u, u_A, u_B, u_C, u_{AB}, u_{BC}	2
AC/BC	u, u_A, u_B, u_C, u_{AC}, u_{BC}	2
A/BC	u, u_A, u_B, u_C, u_{BC}	3
B/AC	u, u_A, u_B, u_C, u_{AC}	3
C/AB	u, u_A, u_B, u_C, u_{AB}	3
A/B/C	u, u_A, u_B, u_C	4

Note that this table does not include models for which one of the three variables is completely omitted. For example, The model A/B is also considered an hierarchical model with terms u, u_A, and u_B, but not terms involving variable C. Models which do not include one of the variables essentially mean that the omitted variable does not contribute to our understanding of the table, and we can collapse the table over this variable to produce, for example, a two-way table involving only variables A and B.

Also note that the degrees of freedom are the number of terms from the saturated model not included in an hierarchical model. For example, model A/BC omits u_{ac}, u_{bc}, and u_{abc} from the saturated model and therefore has three degrees of freedom.

The computations for the term values for hierarchical models are usually not straightforward. Instead, the term values are calculated in an iterative manner. For this reason, a statistical package is a requirement for most hierarchical models.

> Saturated models contain every possible interaction in a table. Hierarchical Models contain fewer terms but are developed in a structured manner. The rule with hierarchical models is that any term in the model automatically means that its higher-order terms are also included. If ABD is in a hierarchical model, then A, B, D, AB, AD, and BD must be in the model. The aim of hierarchical models is to develop a loglinear model that provides good estimates for the table with fewer terms than a saturated model.

Any of these models will provide estimates for the cells in our table. To determine how well a model fits the table, we develop a test statistic as follows:

$$Y^2 = 2 \sum O \log \frac{O}{E} \qquad (31)$$

where O is the observed or actual count in a given table cell and E is the expected count under a particular hierarchical model. The Y' statistic approximately follows a chi-squared distribution.

For tables with three variable (I x J x K), the Y statistic can be written as:

$$Y^2 = 2 \sum_I \sum_J \sum_k x_{ijk} \frac{\log x_{ijk}}{e_{ijk}} \qquad (32)$$

where x_{ijk} is the observed count in cell (i,j,k) and e_{ijk} is the expected count according to the model.

Because the Y^2 statistic follows a chi-quared distribution, we can use Appendix A to test whether a model is appropriate at the 5 percent level. If the calculated Y^2 statistic is less than the entry in Appendix A, then we accept the model as a good fit to the data; conversely, if it is greater, we conclude that our model does not provide a good fit.

We will apple hierarchical models to our example with four dichotomous variables. We will consider the extent of flame damage (fire confined to the room/not confined to the room) as a **response** variable and the other three variables **as explanatory variables.** That is, we want to model how detector performance (Variable B), type of material (Variable C), and involvement of equipment (Variable D) relate to the extent of flame damage (Variable A). Our aim is to select a model which fits the table in a reasonable manner and has fewer terms than the saturated model.

Since variables B, C, and D are explanatory variables, any hierarchical model must include the term BCD, which means any model will also include its higher-order terms (DC, BD, CD, B, C, and D). With these terms as common to all models, we can then concentrate on interactions with our response variable.

Exhibit 6-10 shows the results from several models. Each contains the BCD term along with selected terms involving the response variable. The degrees of freedom for each model is found by counting the number of terms omitted from the saturated model. For example, the first model, AB/ACD/BCD contains all the terms from the saturated model except ABD, ABC, and ABCD. The model therefore has three degrees of freedom.

Only Models 1 and 7 provide a good fit to the data according to our test. Model 1 is defined as AB/ACD/BCD and Model 7 as AB/AD/BCD. We need to determine which of these two models we want to select as the final model. As a general rule, when choices are available, the model to select is the one with the fewest number of terms. Both models contain u_{BCD} and its higher-order effects. In addition, Model I includes u_{AC}, U_{AC}, U_{AD}, and and u_{AB} while Model 7 includes only u_{AB} and u_{AD}. Model 7 is the better choice since it fits the table reasonably well and has two fewer terms.

Exhibit 6-10 Hierarchical Models for Extent of Flame Damage

Chicago Residential Fires-1990

Model	Terms	Y^2	Degrees of Freedom
1	AB/ACD/BCD	2.09*	3
2	AD/ABC/BCD	11.39	3
3	AC/ABD/BCD	10.96	3
4	ACD/BCD	25.97	4
5	AB/AC/AD/BCD	11.56	4
6	AB/AC/BCD	71.62	5
7	AB/AD/BCD	11.57 *	5
8	AC/AD/BCD	36.91	5

Note: An asterisk indicates that the model is significant at the 5 percent level That is. the model produces good estimates of the actual table values.

We could have anticipated that Model 7 would be a good model because of the results in the previous section with the saturated model. In the saturated model, the primary interaction effects were AB, AD, CD, and BCD. With the hierarchical model, BCD is automatically included along with CD as a higher-order effect. The two remaining effects, AR and AD. complete Model 7.

Exhibit 6-11 shows the term values and standardized values for Model 7. All except one of the terms have standardized values with magnitudes greater than 2.0. In particular, all the terms involving variable A in the model are important.

Exhibit 6-11 Extent of Flame Damage Chicago Residential Fires, 1990

Model AB/AD/BCD

Term	Term Value	Standardized Value
Mean	4.807	N/A
A	.577	21.7
B	-.678	-22.4
C	-.484	-17.9
D	-.591	-19.8
AB	.120	4.9
AD	.169	7.7
BC	.008	.3
BD	.084	3.1
CD	-.326	-12.1
BCD	-.084	-3.1

From these term values, we can obtain expected values as shown in Exhibit 6-12 along with the raw data. Most of the expected values are close to the actual data in the table. Ten expected values are within five percent of their actual values and 14 are within ten percent of their actual values. The model therefore provides reasonably accurate estimates for the cells in the table.

In summary, the hierarchical model (AB/AD/BCD) described in Exhibit 6-11 provides an excellent model for the residential fires in Chicago. The model indicates the following key results:

- Containment of a fire to the room of origin is more likely if fire detectors are present and operated.
- Containment of a fire to the room of origin is more likely if equipment is involved in the fire.
- Containment is not related to whether fabric is the type of material ignited.

Exhibit 6-12 Comparison of Expected Values and Actual Table Values

Chicago Residential Fires-1990

A	B	C	D	Actual Table Value	Model Value
1	1	1	1	37	36.5
1	1	1	2	165	163.3
1	1	2	1	218	215.0
1	1	2	2	181	186.3
1	2	1	1	95	109.6
1	2	1	2	504	490.5
1	2	2	1	487	475.9
1	2	2	2	798	808.0
2	1	1	1	6	6.5
2	1	1	2	55	56.8
2	1	2	1	35	38.0
2	1	2	2	70	64.7
2	2	1	1	46	31.4
2	2	1	2	262	275.5
2	2	2	1	125	136.1
2	2	2	2	464	454.0

Key pointers in using loglinear analysis are as follows:
(1) The most important variables for a table are usually variables in a saturated model with standardized values greater than 2.0.
(2) The Y^2 statistic follows the chi-squared distribution. It can therefore be tested to determine whether a model provides a good fit to the table.
(3) The best model for a table is usually the model that provides a good fit according to the Y^2 statistic. If more than one model provides a good fit, you should usually select the model with the fewest number of terms.

We can verify these results by developing two-way tables as shown in Exhibit 6-13. These tables were derived from the figures in Exhibit 6-7. The first table shows that when a detector operated, the odds of fire confinement to the room of origin were 3.62:1 (601 divided by 166). On the other hand, when a fire detector did not operate, the odds drop to 2.10:1 (1,884 divided by 897). Similarly, if equipment is involved in the fire, the odds are 3.95:1 (837 divided by 212) that the fire will be confined to the room of origin, as compared to 1.94: 1 (1,648 divided by 851) if equipment is not involved. Finally, the third table shows similar column

Chapter 6

percentages so that we can conclude independence of extent of flame damage and whether fabric is the type of material ignited.

Exhibit 6-13 Two-Way Interaction Tables

Chicago Residential Fires-1990

	Detector Operated	Detector Did Not Operate	Total
Confined to Room	601	1,884	2,485
Column Percent	78.4	67.7	
Extended Beyond Room	166	897	1,063
Column Percent	21.6	32.3	
Total	767	2,781	3,548
	100.0	100.0	

	Equipment Involved	No Equipment Involved	Total
Confined to Room	837	1,648	2,485
Column Percent	79.8	65.9	
Extended Beyond Room	212	851	1,063
Column Percent	20.2	34.1	
Total	1,049	2,499	3,548
	100.0	100.0	

	Fabric Material	Other Material	Total
Confined to Room	801	1684	2,485
Column Percent	68.5	70.8	
Extended Beyond Room	369	694	1,063
Column Percent	31.5	29.2	
Total	1,170	2,378	3,548
	100.0	100.0	

The key point is that loglinear analysis identified the important and unimportant interactions. We did not have to attempt the identifications by analyzing several two-dimensional and three-dimensional tables. The systematic approach provided by loglinear analysis resulted in an excellent model of the table with important insight into the conditions under which fires are contained to the room of origin.

Summary

Loglinear analysis is an systematic approach for analyzing multi-dimensional tables. The aim of loglinear analysis is to identify the important interactions among the variables in the table. Practical application of loglinear analysis means that the user identifies several different combinations of interactions and then develops a loglinear model for each combination. Each model produces estimates for the cells in the table table these estimates can he compared against the actual table values. The most appropriate model can then he selected as the final representation of the table.

Chapter 6

Chapter 6
PROBLEMS

1. Calculate the odds ratio for the table in Exhibit 6-1 to verify independence of the two variables.

2. Calculate the odds ratios for the three tables in Exhibit 6-13.

3. Use the term values in Exhibit 6-11 to verify the model values in Exhibit 6-12 for cells (1,1,1,1), (1,1,2,2), and (2,1,2,1).

4. In the main example for this chapter, it was shown that fabric as the material ignited was not considered an important variable from the loglinear analysis. As an alternative, we substitute the area of origin to give the following four variables

Variable	Description
A	Fire was confined to the room of origin, or was not
B	Detector performed, or did not (or was not present)
C	Fire started in functional area (assembly area, sleeping room, kitchen, dining area, etc.), or did not
D	Equipment was involved in the fire, or was not

The data for Chicago residential fires for 1990 with these definitions is:

A	B	C	D	Frequency	A	B	C	D	Frequency
1	1	1	1	220	2	1	1	1	25
1	1	1	2	26-1	2	1	1	2	78
1	1	2	1	34	2	1	2	1	16
1	1	2	2	68	2	1	2	2	47
1	2	1	2	433	2	2	1	1	119
1	2	1	2	764	2	2	1	2	363
1	2	2	1	147	2	2	2	1	56
1	2	2	2	445	2	2	2	2	329

(continued from question 4)

The saturated model gives the following results:

Term	Term Value	Standardized Value	Term	Term Value	Standardized Value
A	.486	16.2	BD	.094	3.1
B	-.707	-23.6	CD	.101	3.4
C	.413	13.8			
D	-.475	-15.8	ABC	.094	3.1
			ABD	.007	.2
AB	.074	2.5	ACD	.029	1.0
AC	.190	6.3	BCD	-.047	-1.6
AD	.157	5.2			
BC	.105	3.5	ABCD	.042	1.4

a. Analyze the standardized values to determine the most important interactions among the variables.

b. Treat variable A as the response variable and develop several hierarchical models of interest.

5. For the data from Problem 4, the following are model results for selected hierarchical models.

Model	Terms	Y^2	Freedom
1	ABC/AD/BC/BD/CD	3.5	4
2	AB/AC/AD/BC/BD/CD	10.9	5
3	ABC/BC/BD/CD	54.8	5
4	AB/AC/BC/BD/CD	62.4	6
5	AB/AD/BC/BD/CD	52.2	6
6	AC/AD/BC/BD/CD	28.5	6

a. Select the two best models from these alternatives.

b. Which of the two models is the better model?

c. From the data in Problem 4, develop two-way tables for variable A against the other three variables.

d. Calculate the odds and odds ratios for each of the three tables.

Chapter 6: Problems

CORRELATION AND REGRESSION

Introduction

In this chapter we present correlation and regression analysis for continuous data. Correlation is a statistical measure which indicates the degree to which one variable changes with another variable. We know, for example, that calls from citizens for Emergency Medical Services (EMS) increase with population growth. That is, as population increases, we normally expect more medical services calls from citizens. Statistically, we say there is a positive correlation between population and EMS calls. The correlation measures the strength of association between the two variables.

With regression we go a step further and create a straight line, with an associated regression equation, through a group of points. The result is as an analytical description of the relationship between two variables. The regression equation defines the straight line algebraically. As we will see in this chapter, population predicts fire department workload fairly accurately with a regression equation. In general, we can say that a regression line summarizes a group of points in a manner similar to the way an average summarizes a group of numbers.

This chapter starts with the scatter diagrams illustrated in Chapter 3 and proceeds with the calculation of correlation. Next, we give an example of a regression equation and discuss several applications. We then discuss how to calculate a regression line. The chapter concludes with a second example on the relationship between increases in population and increases in EMS activity for a fire department.

Scatter Diagram

Exhibit 7-1 shows the data for a scatter diagram presented in Chapter 3 on population protected and number of fires during 1989 for 18 selected jurisdictions." Exhibit 7-2 is a scatter diagram of the data. The horizontal axis gives population (in thousands) and the vertical axis gives the number of fires. We can see from the exhibit that fire levels are higher with greater population. The general trend is clear although the pattern is not perfect. We use the term "not perfect" to mean that the points do not fall on a straight line.

With relationships depicted in this manner, the usual terminology is to label one variable as the **independent** variable, and the other as the

---

9. In Chapter 3, we identified Houston, Texas and Detroit, Michigan as outliers in the data That is, they have a different relationship than the other cities between population and EMS calls For the purposes of this chapter, we have therefore dropped them from the analysis.

Exhibit 7-1 Population and Fires for 1989–Selected Cities

City	Population Protected	Fires
Arlington, Texas	254,500	1,644
Wichita, Kansas	261,000	1,978
St. Paul, Minnesota	264,800	2,041
Corpus Christi, Texas	274,500	1,769
Newark, New Jersey	275,200	4,442
Norfolk, Virginia	280,000	2,140
Toledo, Ohio	354,600	3,597
Minneapolis, Minnesota	356,700	2,897
Omaha, Nebraska	360,000	2,336
Cincinnati, Ohio	364,000	2,645
Fort Worth. Texas	450,100	5,075
Denver, Colorado	500,000	4,244
Cleveland, Ohio	505,600	6,324
Boston, Massachusetts	574,300	6,479
El Paso, Texas	603,900	4,333
Columbus, Ohio	660,000	4,561
Dallas, Texas	982,800	10,210
San Antonio, Texas	956,200	8,957

dependent variable. In Exhibit 7-2, population serves as the independent variable and fires as the dependent variable. The independent variable is viewed as influencing the dependent variable. Obviously, population influences the number of fires; greater population means more structures and more vehicles, which, in turn, may lead to more fires. The question is how strong is the relationship between population and fires.

Exhibit 7-2 Scatter Diagram of Population & Fires-Selected Cities-1989

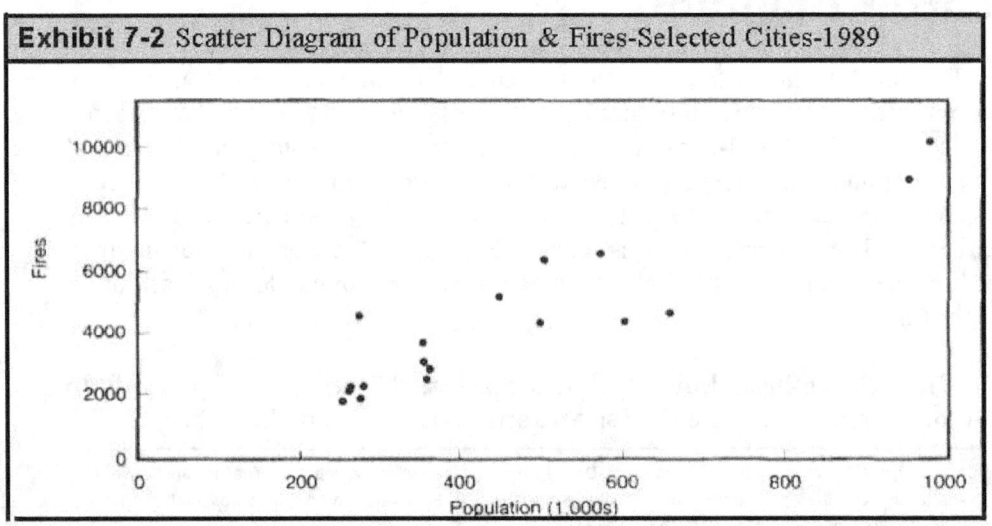

Correlation measures the strength of association between two variables. One variable is usually called the **independent variable** and the other is called the dependent variable. A *strong association* between the two variables means that knowing the value of the independen Variable helps to predict the value of the dependent variable. conversely, a weak association means that the independent variable does not help much in determining the values of the dependent variable.

Correlation Coefficient

The **correlation coefficient,** more commonly just called **correlation,** measures the strength of association between two variables. In the next section, we will provide the calculation for the correlation coefficient, but here we want to understand it and discuss its key properties.

The first property to know is that a correlation is always between -1 and +1. A correlation of exactly -1 or +1 is called a perfect correlation, and means that all the points fall on a straight line. If a correlation is zero, then there is no association between the two variables. That is, one variable does not assist in knowing or predicting the value of the second variable. As a correlation moves from zero towards either +1 or -1, the strength of association between the two variables also increases.

Exhibit 7-3 shows several scatter diagrams with correlations ranging from 0 to +l. Note that when the correlation is zero, the dots are randomly scattered with no pattern. As the correlations increase, the association becomes stronger--the dots begin to cluster together and we can start to visualize a straight line through the points.

The correlations in Exhibit 7-3 are positive because the patterns of the dots always move from the lower left to the upper right. Just the opposite is true with negative correlations, as shown in Exhibit 7-4, where the patterns move from the upper left to the lower right. The **negative** relationship means that as the independent variable **increases,** the dependent variable **decreases.** The direction of the dots is the distinguishing feature between negative and positive correlations. A correlation of -.9 is just a strong as a correlation of +.9, but the pattern of the dots is in the opposite direction.

A correlation has no units of measurement associated with it. That is, a correlation is not expressed in terms of the independent or dependent variable. It is dimensionless. The interpretation of a correlation coefficient generally depends on its closeness to -1, 0, or +1. Correlations close to 0 mean there is no association between the two variables while correlations close to -1 or +1 indicate strong associations.

Chapter 7

Exhibit 7-3 Positive Correlations

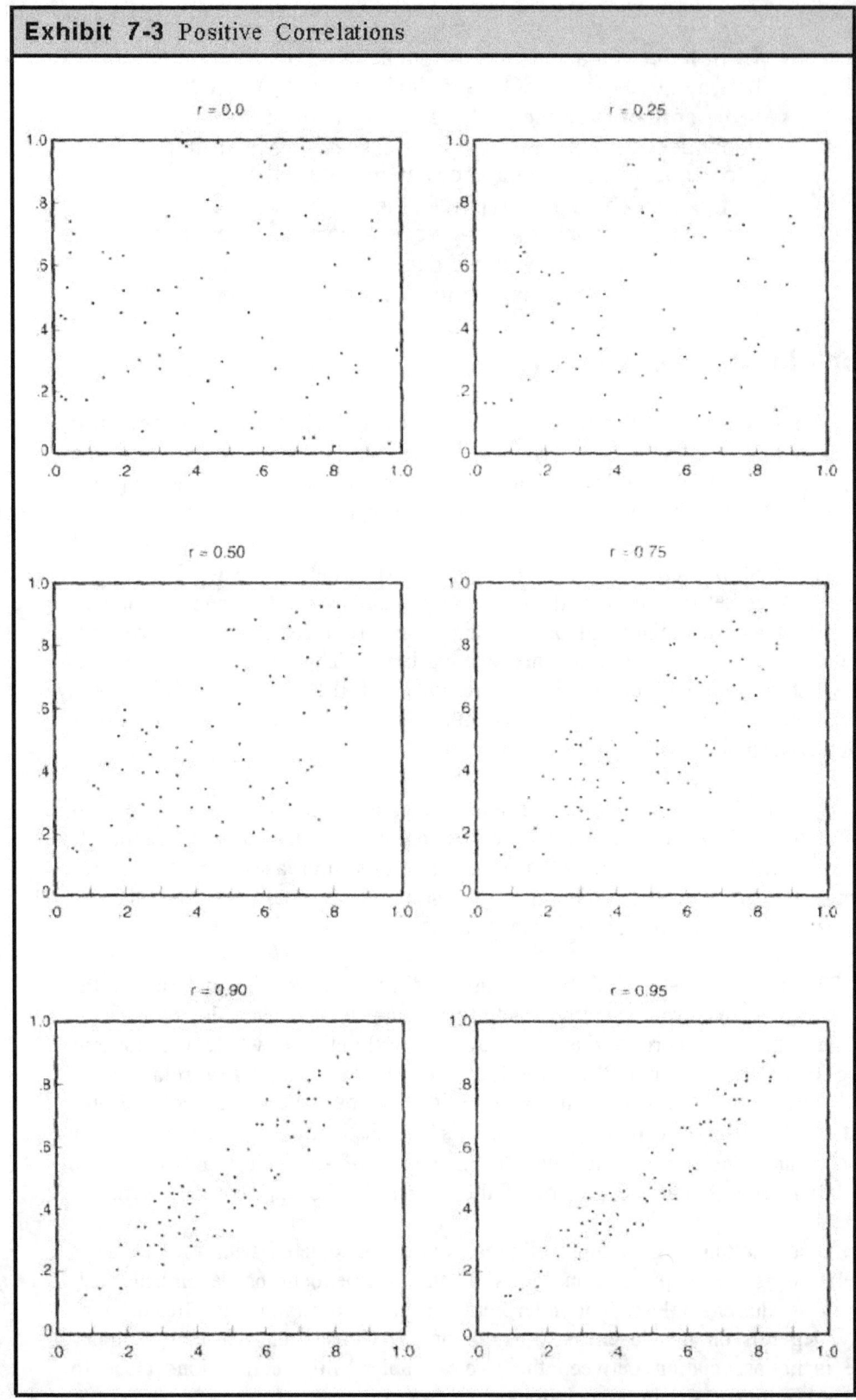

Exhibit 7-4 Negative Correlations

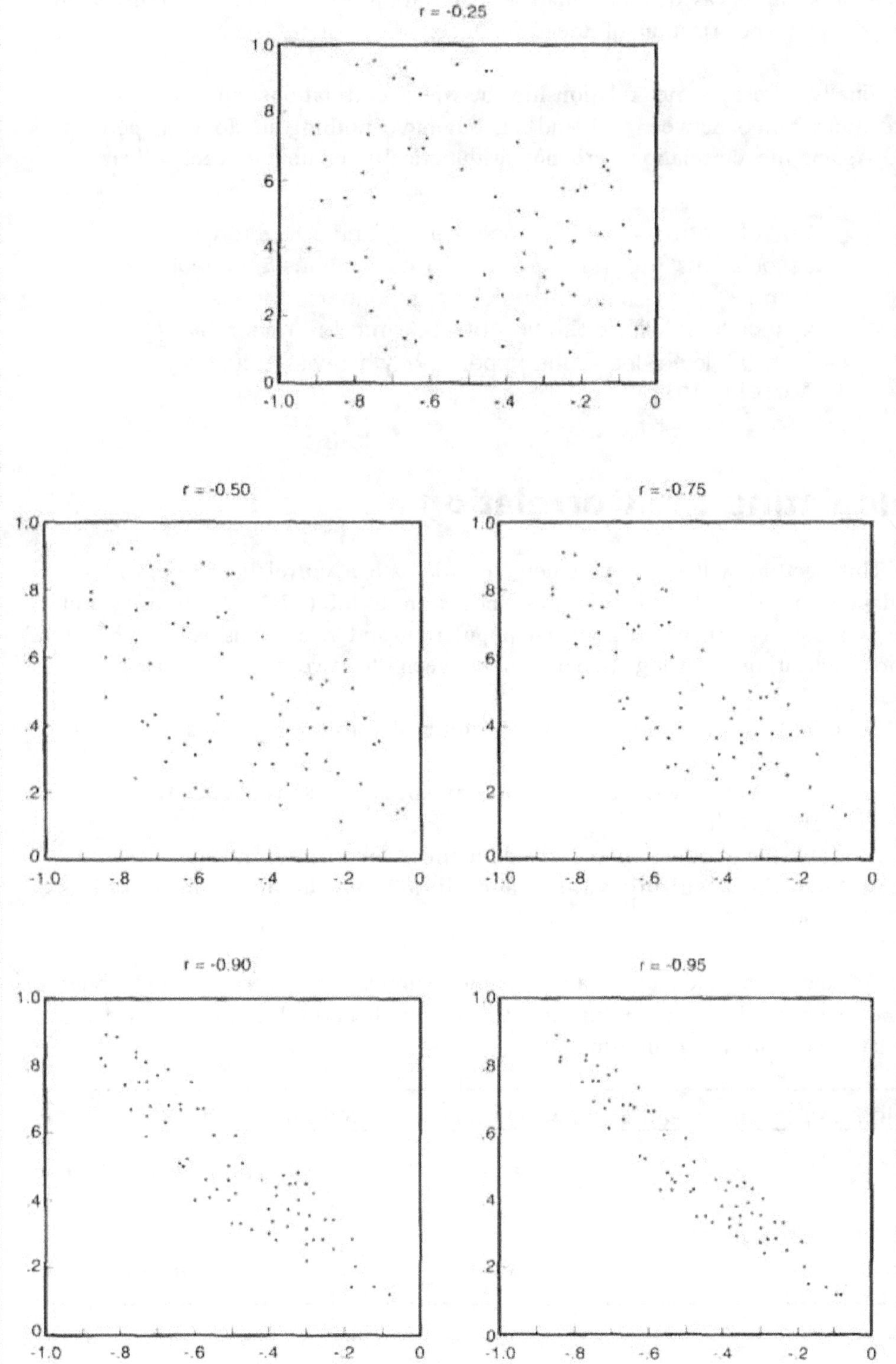

r = -0.25

r = -0.50

r = -0.75

r = -0.90

r = -0.95

Chapter 7

Another important point to know is that correlations are **not** arithmetically related to each other. For example, a correlation of .6 is not twice as strong as a correlation of .3. We can obviously say that a correlation of .6 reflects a stronger association than a correlation of .3, but we must stop short of exact specification of the difference.

Finally, there is no relationship between correlations and percentages. Correlations range between - 1 and +1, but have nothing to do with percentages. Again, the correlations are not arithmetically related to each other.

> **Correlations** are always between -1 and +1. Stronger associations are indicated as the correlations approach -1 and +1 while correlations close to 0 indicate no association. With **positive correlations,** the dots move from the lower left to the upper right while with **negative correlations,** the opposite is true.

Calculating the Correlation

This section will show you how to calculate a correlation. We will use the data from the selected cities, as shown in Exhibit 7-1. As it turns out, there is a correlation of .92 between population and fires. This is a high correlation indicating a strong association between the two variables.

To calculate the correlation, we perform the following steps:

1. Convert the values of both variables to **standard units,** as defined below.
2. Take the product of the standard units for each pair.
3. Sum the resulting values and divide by the number of points minus 1.

To convert a value into standard units, subtract the average'" and divide by the standard deviation. Returning to the numbers in Exhibit 7-1, we can calculate the following information:

Exhibit 7-5 Averages and Standard Deviations for Selected Cities		
	Average	Standard Deviation
Population (in thousand)	459.90	224.78
Fires	4,204.00	2,469.43

10. In this chapter, the average is the arithmetic mean (rather than the median or mode) and the standard deviation is the sample standard deviation (rather than the population standard deviation).

$$\text{Standard units} = \frac{x_j - \bar{x}}{s}$$

Where $\bar{x}$ = arithmetic mean

s = sample standard deviation

Exhibit 7-6 gives the steps for the correlation calculation. The column labeled "Population Standard Units" lists the standard units for the population of each city. For the first population of 254.5, we obtain the standard unit by subtracting the average of 459.9 and dividing by the standard deviation of 218.45. The calculation proceeds as follows:

$$\text{Standard Unit} \atop \text{(for 254.5)} = \frac{254.4 - 459.9}{224.78} = \frac{-205.4}{224.78} = -0.91$$

For the standard unit for the first figure for fires, the calculation proceeds in the same manner using the average and standard deviation for fires.

$$\text{Standard Unit} \atop \text{(for 1,644)} = \frac{1,644 - 4,204}{2,469.43} = \frac{-2,560}{2,469.43} = -1.04$$

Exhibit 7-6 Correlation Calculation

City	Population (thousands)	Fires	Population Standard Units	Fires Standard Units	Product
Arlington	254.5	1,644	-0.91	-1.04	0.95
Wichita	261.0	1,978	-0.88	-0.90	0.80
St. Paul	264.8	2,041	-0.87	-0.88	0.76
Corpus Christi	274.5	1,769	-0.82	-0.99	0.81
Newark	275.2	4,442	-0.62	0.10	-0.08
Norfolk	280.0	2,140	-0.80	-0.84	0.67
Toledo	354.6	3,597	-0.47	-0.25	0.12
Minneapolis	356.7	2,897	-0.46	-0.53	0.24
Omaha	360.0	2,336	-0.44	-0.76	0.34
Cincinnati	364.0	2,645	-0.43	-0.63	0.27
Ft. Worth	450.1	5,075	-0.04	0.35	-0.02
Denver	500.0	4,244	0.18	0.02	0.00
Cleveland	505.6	6,324	0.20	0.86	0.17
Boston	574.3	6,479	0.51	0.92	0.47
El Paso	603.9	4,333	0.64	0.05	0.03
Columbus	660.0	4,561	0.89	0.14	0.13
Dallas	982.8	10,210	2.33	2.43	5.66
San Antonio	956.2	8,957	2.21	1.92	4.25
Total					15.57

Correlation (Total divided by 17) = .916

The last column gives the product of the two calculations, which is 0.95 (-0.91 times -1.04). These calculations are performed for each of the 18 data points. The sum of this column is 15.57, and the correlation is defined as this total divided by 17, which results in the correlation of .916.

Other Ways to Calculate Correlation*

Several equivalent formulas exist for calculating correlations. We selected the above procedure to illustrate the connections between correlations, averages, and standard deviations. The correlation is an average of the products of standard units, except that we divided our sum by the number of points minus one.

In algebraic terms, the correlation can be seen on the following page.

$$\text{Correlation} = \frac{\sum (x_i - \bar{x})(y_i - y)}{(n-1)\, s_x\, s_y}$$

where s_x is the sample standard deviation of x s_y is the sample standard deviation of y. This is the equation that was used to Calculate the correlation in Exhibit 7-6.

An equivalent formulation for the correlation is given by the following:

$$r = \frac{\sum (x_i - \bar{x})(y_i - \bar{y})}{\sqrt{\sum (x_i - \bar{x})^2 \sum (y_i - \bar{y})^2}}$$

In this equation, we have taken the prior equation and replaced the standard deviations with their actual formulas.

Another way of expressing the correlation is by the following:

$$r = \frac{\text{Cov}(x,y)}{\sqrt{\text{Var}(x)\,\text{Var}(y)}} \quad \text{where Cov}(x,y) = \frac{\sum (x_i - \bar{x})(y_i - \bar{y})}{n-1}$$

Var(x) is the sample variance of x (the square of the sample standard deviation for x), and Var(y) is the sample varance of y.

The quantity Cov(x,y), which is called the covariance of x and y, measures the extent to which the two variables rise and fall together. By itself, the covariance is hard to interpret because it is not expressed in units of either x or y and it does not have an upper or lower bound like a correlation. However, dividing the convariance by the two variances standardizes the final result to a correlation, which is always between -1 and +1.

Exhibit 7-7 Correlation Matrix

Variable	Population	Structure Fires	Vehicle Fires	Other Fires	Civilian Injuries	Fire Civilian Fatalities	Fire Service Injuries	Service Fatalities	Dollar Loss
Population	1.00	.82	.73	.94	.25	.42	.53	.23	.37
Structure Fires	.82	1.00	.90	.79	.32	.70	.68	-.32	.53
Vehicle Fires	.73	.90	1.00	.79	.18	.86	.77	-.24	.30
Other Fires	.94	.79	.80	1.00	.26	.53	.53	-.22	.29
Civilian Injuries	.25	.32	.18	.26	1.00	.14	.31	-.26	.76
Civilian Fatalities	.42	.70	.86	.53	.14	1.00	.66	-.24	.17
Fire Service Injuries	.53	.68	.77	.53	.31	.66	1.00	-.27	.33
Fire Service Fatalities	-.23	-.32	-.24	-.22	-.26	-.24	-.27	1.00	-.18
Dollar Loss	.37	.53	.30	.29	.76	.17	.33	-.18	1.00

Correlation Matrix*

Exhibit 7-7 on the previous page shows **a correlation matrix** for several variables for the 18 cities discussed in the previous section. Each entry in the matrix is a correlation. For example, the first line shows a correlation of .82 between population and structure fires.

Note that the diagonal of the correlation matrix is always 1.00, since this is the correlation of a variable with itself. Also, the correlation is symmetric about the diagonal. The correlation of .73 (between population and vehicle fires) from the first line also appears in the first column. The lower half of the matrix could, in fact, be omitted without losing any information about the correlations. However, the complete table is usually displayed co you can find specific correlations easier.

There are several high correlations in the matrix. For example, the correlation between population and structure fires is .82, and between population and other fires is .94. On the other hand, there are several low correlations, indicating relatively little relationship between the variables. The correlation between population and civilian injuries is only .25. This means that population does not necessarily provide a good indicator of the number of civilian injuries from fires.

Regression Line

Returning to Exhibit 7-2, We now want to know more about a straight line that best fits the dots. As we will show in the next section, the **regression line** for relating population to fires for these cities is given by:

Fires = 10.06 x Population -424.03

The value 10.06 is called the **regression coefficient,** or **slope,** of the regression line, and the value -424.03 is the **constant** or **intercept.** The next section explains how to calculate these values.

Exhibit 7-8 shows the regression line within the scatter diagram for population and fires. Note that the dots cluster nicely around the regression line. The regression line is a representation of these dots, just as the average is a representation of a single list of numbers.

> The regression line estimates the average value for the dependent variable for a given value of the independent variable. the regression line is the numerical representation of a scatter diagram.

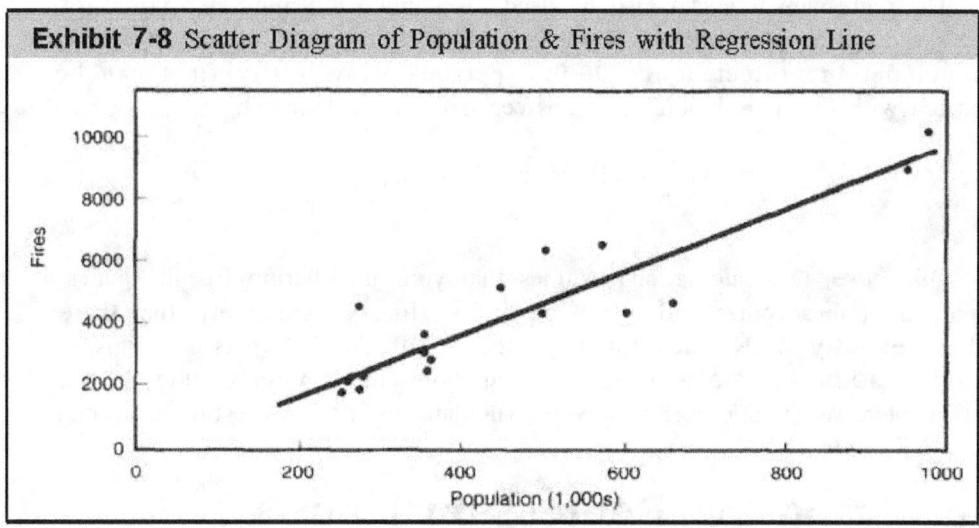

Exhibit 7-8 Scatter Diagram of Population & Fires with Regression Line

To see how to use this regression, consider the population of 500,000 for Denver, Colorado. The number of fires estimated by the regression line is then:[11]

$$\text{Denver Fires} = (10.06 \times 500.0) - 424.03$$

$$= 5,030 - 424.03$$

$$= 4,607.5$$

Denver actually experienced 4,244 calls. The regression estimate was off by 363.5 fires, or about 8.5 percent.

From the equation, you can see that the number of fires changes by 10.06 every time the population variable changes by 1 (that is, the population changes by 1,000). In general, the regression coefficient reflects the change in the dependent variable with a one unit change in the independent variable.

Another feature of a regression line is that the line always goes through the point created by the averages for the two variables. In our example, the average population (in thousands) is 459.90 (see Exhibit 7-5). For this population average of 459.90, we have the following from the regression line:

$$\text{Denver Fires} = (10.06 \times \text{Population}) - 424.03$$

$$= (10.06 \times 459.90) - 424.03$$

$$= 4,204.0$$

The result is **4,204.0** fires, which is the average number of fires for the 18 cities.

---

11. If you duplicate this calculation, you will not get an answer of exactly 4,607.5 fires because of rounding errors. More precisely, the slope is 10.0613 and the intercept is -424.0344.

Chapter 7

The regression line can also be used to estimate the number of fires for cities with other populations. Suppose, for example, that your jurisdiction has population protected of 700,000 persons. How many fires can be expected with this population? The answer is easily, calculated:

$$\text{Fires} = (10.06 \times 700) - 424.03$$
$$= 6,618$$

Of course, this calculation assumes that your jurisdiction fits the general pattern of these cities and is not an outlier. It is also unlikely, that there will be **exactly** 6,618 fires for a population of 700,000 persons. This is strictly **a point estimate** obtained by applying the regression line. In the next chapter, **we** will describe how to calculate an **interval** estimate around this point estimate.

Calculating the Regression Line

The regression line has the general form:

$$y = mx + b$$

Where y is the dependent variable, x is the independent variable, m is the slope (or regression coefficient), and b is the intercept (or constant). In our example, the dependent variable is the number of fires and the independent variable is population. We will now calculate the slope and intercept for our example. Just as there are several equivalent equations for correlation, there are also several ways to calculate the slope and intercept, all of which result in the same answers. Our approach takes advantage of the information we have about the averages, standard deviations, and correlation for our two variables.

In Exhibit 7-9, we have placed a dot at the pair formed by the two averages (459.9 population and 4,204 fires). As stated in the previous section, we want the regression line to pass through this point. We get a second dot by moving one standard deviation for the population to the right and upward by one standard deviation for fires **times** the correlation. The regression line is then formed graphically by drawing a straight line through these two points.

Algebraically, the slope of the straight line is:

$$\text{Slope} = \frac{r \times \text{S.D. of Fires}}{\text{S.D. of Population}} = \frac{(.916 \times 2,469.43)}{224.78} = 10.06$$

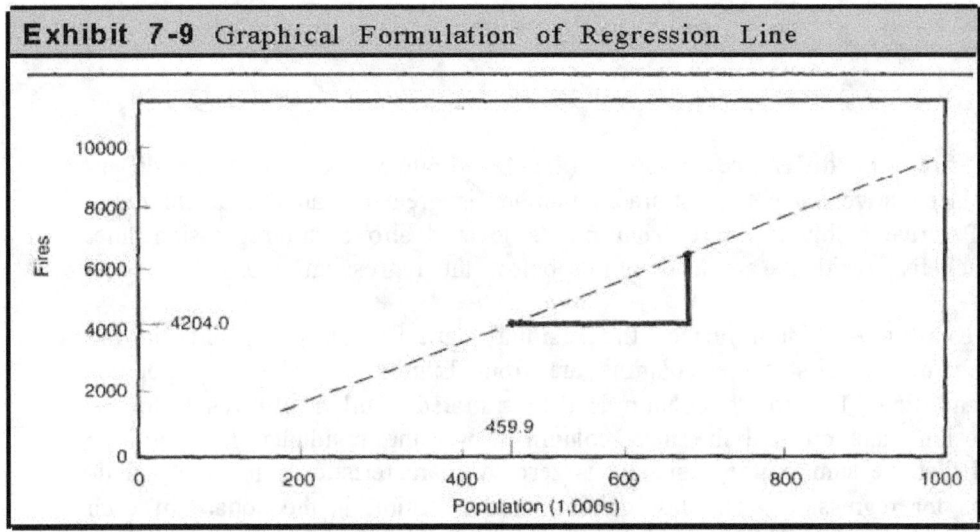

Exhibit 7-9 Graphical Formulation of Regression Line

Calculating the intercept takes advantage of the fact that the regression line goes through the point created by the two averages. That is, we determine the intercept, b, by:

Fires Average = m x Population Average + b

4,204.O = 10.06 x 459.90 + b

-424.03 = b

The final result is the regression line:"

Fires = 10.06 x Population -424.03

> In a **regression line,** when the independent variable Changes by one standard deviation , the dependent variable changes by **r** standard deviations.

Standard Error of the Regression*

The **standard error** of the regression is an estimate of the accuracy of the regression line you have developed. In this section we will present two equivalent ways to calculate the standard error. One way revolves around the calculation of the **residuals** of the regression line, while the other way is a quicker algebraic formula.

The residuals for a regression are the differences between the data points and the estimates from the regression equation. As we showed earlier, the population protected for Denver was 500,000 so that the estimated number of fires is 4,607.5 (10.06 x 500 - 424.03). Since Denver actually had

12 The more precise numbers were used in the calculations to get these results

4,244 fires, the difference is -363.5 (4,244 - 4,606.5). Note that the difference is negative since the estimated number is greater than the actual number. Positive residuals come from points located above the regression line, and negative residuals are from points below the regression line.

Exhibit 7-10 summarizes the residual data for all the points in the regression. The first three columns are from Exhibit 7-1 she-wing population and fires. The fourth column is the estimated number of fires from the regression equation and the next column shows the residuals. It should be noted that the sum of the residuals is zero, a characteristic of residual calculations for regressions. The last column in the exhibit is the square of each residual (the residual multiplied by itself). The total for this last column is 16,681,736.4, which is sometimes referred to as the **sum of squared errors, or SSE.**

Exhibit 7-10 Residuals for Regression Line

City	Population (thousands)	Fires	Estimated Fires	Residual	Residual Squared
Arlington	254.5	1,644	2137.0	-493.0	243,081.3
Wichita	261.0	1,978	2,202.4	-224.4	50,374.7
St. Paul	264.8	2,041	2,240.7	-199.7	39,873.3
Corpus Christi	274.5	1,769	2,338.3	-569.3	324,097.2
Newark	275.2	4,442	2,345.3	2,096.7	4,395,985.1
Norfolk	280.0	2,140	2,393.6	-253.6	64,334.6
Toledo	354.6	3,597	3,144.4	452.6	204,890.0
Minneapolis	356.7	2,897	3,165.5	-268.5	72,084.1
Omaha	360.0	2,336	3,198.7	-862.7	744,239.4
Cincinnati	364.0	2,645	3,238.9	-593.9	352,771.4
Ft. Worth	450.1	5,075	4,105.4	969.6	940,160.4
Denver	500.0	4,244	4,607.5	-363.5	132,155.2
Cleveland	505.6	6,324	4,663.9	1660.1	2,755,981.4
Boston	574.3	6,479	5,355.2	1,123.8	1,262,876.3
El Paso	603.9	4,333	5,653.1	-1,320.1	1,742,640.2
Columbus	660.0	4,561	6,217.6	-1656.6	2,744,431.9
Dallas	982.8	10,210	9,466.0	744.0	553,518.6
San Antonio	956.2	8,957	9,198.3	-241.3	58,241.3
Total				0.0	16,681,736.4

We calculate the standard error by the following equation:

$$\text{Standard Error} = \sqrt{\frac{SSE}{n-2}}$$

where SSE is the Sum of Squared Errors

In this equation, n is the number of points. In our example, we have 18 points, so that n-2 = 16. The mean square error is therefore calculated as:

$$\text{Standard Error} = \sqrt{\frac{16,681,736.4}{16}} = \sqrt{1,042,608.5} = 1,021.08$$

The standard error tells us how far, on average, the estimates from the regression line deviate from the actual numbers. A small standard error reflects a good fit of the regression line to the data while a large standard error means the regression line is not very representative of the data points. A useful rule of thumb is to see if the standard error is small relative to the dependent variable. In our example, the number of fires ranges between 1,644 and 9,466 with an average of 4,204.0 calls. Our standard error of 1,021.08 is relatively small compared to these data values.

The standard error has another interesting feature. From statistical theory, approximately 68 percent of the actual values should be within one standard error and 95 percent within two standard errors. In our example, 13 of the 18 cities have fire figures within one standard error and 17 are within two standard errors. Thus, these results are in line with statistical theory.

The calculation of the standard error in the above manner obviously is a time consuming job when you have a large number of points. A more direct way takes advantage of knowing the correlation coefficient and standard deviation of the dependent variable, as reflected in the following equation:

$$\text{Standard Error} = \sqrt{\frac{n-1}{n-2}} \times \text{S.D. of Fires} \times \sqrt{1 - r^2}$$

Since the standard deviation for the fires is 2,469.43 and the correlation is .916, the standard error can therefore be calculated as:

$$\text{Standard Error} = \sqrt{\frac{n-1}{n-2}} \times \text{S.D. of Fires} \times \sqrt{1 - r^2}$$

$$= \sqrt{\frac{17}{16}} \times 2,469.43 \times \sqrt{.161}$$

$$= 1.031 \times 2,469.43 \times .401$$

$$= 1,021.08$$

The advantage of this equation is that we avoid the arduous residual calculations by knowing the correlation and standard deviations.

Chapter 7

Coefficient of Determination: Explained Variation*

The **coefficient of determination** is another way to determine how well a regression line fits the data points. The coefficient of determination is motivated by the following argument. One crude and simple approach for predicting a variable is merely to use the average as the prediction. With this approach, the same prediction (namely, the average) is always made regardless of the value of independent variable. We can then form squared residuals from the average just as we did in the previous section. The result is called the **total sum of squares,** or **SST:**

$$SST = \sum (y - \bar{y})^2$$

where y is the actual number of fires for a city and $\bar{y}$ is the average number of fires.

In our example, the SST is 103,667,214 . We should note that the sample variance is the SST divided by 17. Thus, we can view SST as a measure of variability in the fire figures.

The coefficient of determination is detined as:

$$R^2 = \frac{SST - SSE}{SST}$$

where SST is the total sum of squares and SSE is the sum of squares for errors.

You may recall that SSE is derived from the residuals obtained as the difference between the actual values and the predicted values. In our example, we calculated the SSE (see Exhibit 7-10) as 2,311,594.1. The coefficient of determination for our example is therefore:

$$R^2 = \frac{103,667,214 - 16,681.736}{103,667,214} = .839$$

We can interpret R^2 in the following way. With the regression equation, the amount of error in the predictions (as measured by the sum of squared errors) is 83.9 percent smaller than when the average is used as the predictor. The value R^2 therefore indicates how much better the linear regression equation is over simply using the average.

Another way to view R^2 is to say that population explains 83.9 percent of the variability in the fire figures. Or conversely, 83.9 percent of the variability in the fire figures is explained by population.

An important feature of the coefficient of determination is that it is equal to the square of the correlation coefficient. In our example, the correlation is .916 and the square of this number is .839. The designation of the coefficient of determination by R^2 is intentional to indicate its relationship to the correlation coefficient. r.

In summary, if we know the correlation coefficient, then a quick calculation shows how much variation will be explained by a regression line. In Exhibit 7-7, we showed correlations for several other variables. The exhibit indicated a correlation of .25 between population and civilian injuries. If we performed a regression of population against civilian injuries, we would explain only 6.25 percent of the variation (since the square of the correlation is .0625 or 6.25 percent). The point is that we could perform such a regression, but the results would not be very satisfying.

An Example with Population and EMS Calls

In this section we present another regression example showing the relationship between population and EMS calls in Prince William County, Virginia. The Prince William County Fire Department has 15 fire stations, of which 4 stations include paramedics for handling EMS calls. Exhibit 7-11 shows the population growth in the county from 1981 to 1991 along with the number of EMS calls handled by paramedics. (You may recall that we showed a scatter diagram of these data in Chapter 3.) One immediate conclusion is that the county is growing as reflected by the population increase of 71,600 over the eleven-year period. The number of EMS calls also increased substantially over the same time period from 9,538 calls in 1981 to 12,744 calls in 1991.

Exhibit 7-11 Population & EMS Calls Prince William County, Virginia

Year	Population	EMS Calls
1981	152,300	9,538
1982	156,700	9,578
1983	159,200	9,657
1984	164,100	9,744
1985	169,700	10,072
1986	176,000	11,703
1987	184,700	11,982
1988	205,000	11,843
1989	221,300	13,074
1990	219,000	13,175
1991	223,900	12,744

Chapter 7

Exhibit 7-12 gives the basic statistics and regression line for population and EMS calls in the county.

Exhibit 7-12 Basic Statistics and Regression Results

Variable	Average	Standard Deviation
Population (thousands)	184.72	27.70
EMS Calls	11,191.82	1,491.31

Correlation (r): .946
Coefficient of Determination (R^2): .895
Regression line: EMS Calls = 50.96 x Population + 1,778.6
Standard Error: 506.8

Exhibit 7-13 shows a scatter diagram of our data along with the regression line. In this example, the fit of the regression line is excellent. An application of this regression line is to anticipate the number of EMS calls for future years. Suppose the county expects the population to reach 300,000 residents in the next few years. How many EMS call can be expected with this population? A point estimate can be made with the regression line:

$$\text{EMS Calls} = (50.96 \times 300) + 1{,}778.6$$
$$= 17{,}067$$

The implication of this increase is that the fire department probably needs to start considering expansion of its EMS program to handle the increased workload.

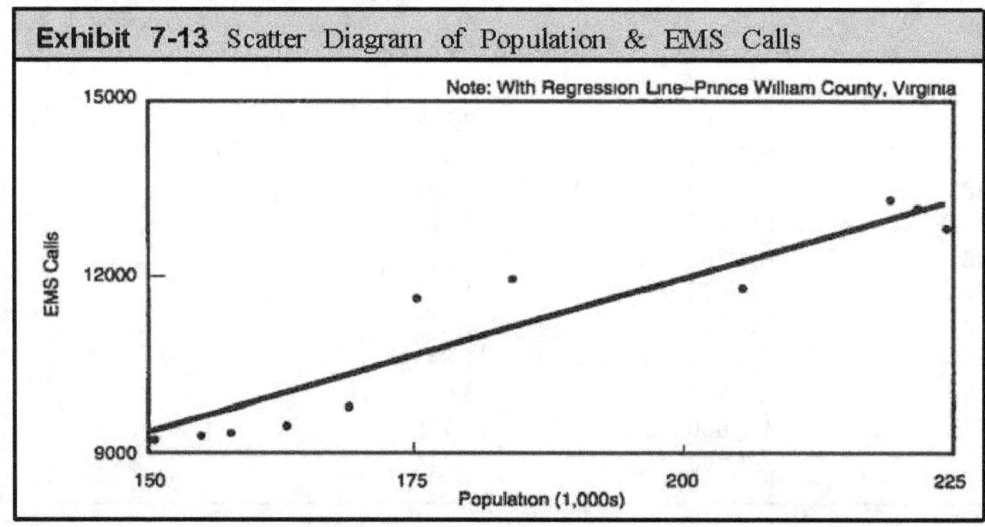

Exhibit 7-13 Scatter Diagram of Population & EMS Calls

Summary

Correlation and regression analysis are two powerful tools for analyzing the relationship between two continuous variables. A correlation close to zero tells you there is no relationship, which means that knowing the value of one of the variables does not tell you much about the value of the other variable. Correlations close to -1 or +1 indicate a strong relationship between two variables. We found, for example, a high correlation between population of jurisdictions and the number of fires. Regression analysis provides a way to quantify the relationship of two variables. The regression line is a representation of the scatter diagram of the two variables. We can apply the regression line to estimate the value of one variable given the other variable. With the regression of population on fires, for example, we can make estimates on the number of fires that a jurisdiction can expect given its particular population.

Chapter 7

PROBLEMS

1. Do the following for each of the figures below

a. Estimate the correlation for each figure.

b. The second figure has a clear outlier. What is the correlation without this outlier.

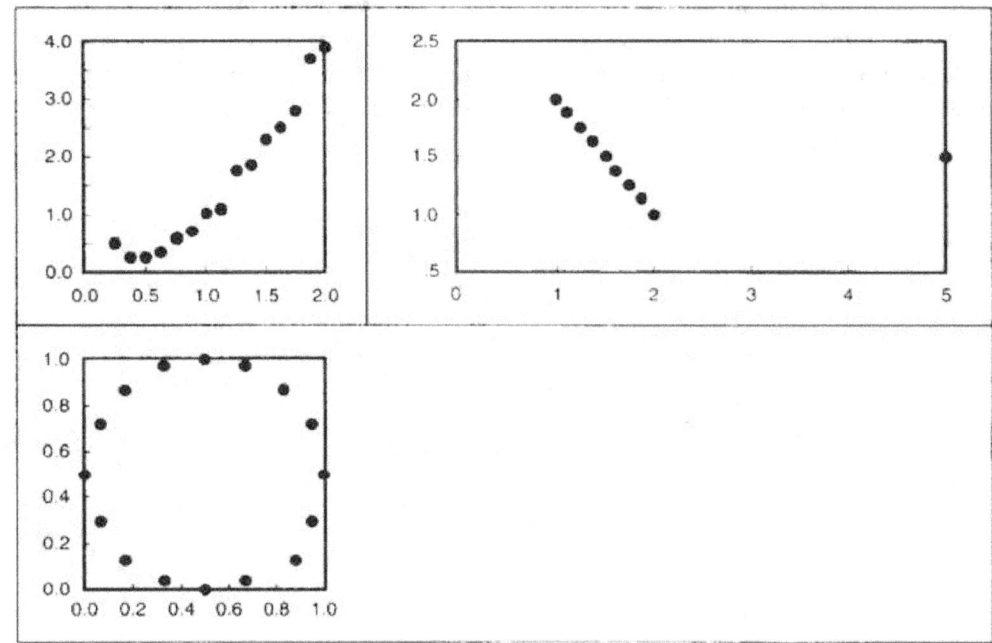

2. The following data, which are from Memphis, Tennessee for 1991, are for structure fires in which the cause of the fire was children playing. All the fires in this list had incident times less than 30 minutes (incident time is from time of dispatch to the scene until time of completion). Develop a scatter diagram and estimate the correlation between incident time and dollar loss for these fires.

Incident Time	Dollar Loss	Incident Time	Dollar Loss
15	700	12	400
18	3,000	16	800
22	250	26	1,000
26	35	22	1,500
14	100	17	5,000
12	500	18	550
28	150	27	250
16	300	11	50

3. The department responded to an additional 73 structure fires caused by children playing in which the incident times were greater than 30 minutes. For these fires the correlation between incident times and dollar losses was .84, and the regression equation relating losses to incident time was as follows:

Fire Loss = 206.2 x Incident Time - 6,266.4

a. Calculate the estimated fire loss for an incident time of 45 minutes and an incident time of 3 hours (180 minutes).

b According to this regression, how much will fire loss increase when incident time increases 10 minutes?

c. In general, fire losses and incident times appear to have a fairly high correlation for incidents requiring more than 30 minutes, but not for incidents less than 30 minutes. Give reasons as to why this result might be true.

4. The number of fires in most jurisdictions has decreased over the last 20 years. However, the following data shows that the total dollar loss due to fires has steadily increased in the United States.

Year	Total Fire Losses (in millions)	Year	Total Fire Losses (in millions)
1967	$1,707	1979	$4,851
1968	1,830	1980	5,579
1969	1,952	1981	5,625
1970	2,328	1982	5,894
1971	2,316	1983	6,320
1972	2,304	1984	7,602
1973	2,639	1985	7,753
1974	3,190	1986	8,488
1975	3,190	1987	8,634
1976	3,558	1988	9,626
1977	3,764	1989	10,210
1978	4,008		

Source: Insurance Information Institute, New York, New York. Insurance *Facts,* annual publications.

a. Develop a scatter diagram showing the year and total dollar loss.

b. The last two digits of the year (67, 68, 69, etc.) can be treated as a variable and we can then develop a regression equation between year and total dollar loss. The correlation between year and dollar loss is very high at .9753, and the means and standard deviations of these variables are as follows:

Variable	Mean	Standard Deviation
Year	78	6.782
Total Fire Losses	4,929.043	2,696.394

Determine the regression line of total fire losses as the dependent variable and year as the dependent variable.

c. With your regression equation, estimate the total fire losses for 1988, 1989, and 1990.

d. Draw the regression line on your scatter diagram. The standard error for this regression equation is 609.06. Draw a line above and below the regression line representing the standard error.

MULTIPLE REGRESSION

Introduction

Chapter 7 discussed regression with only one independent variable (population). We were able to develop a reliable regression line because a high correlation existed between population and fires in cities. In other situations, however, we may know that several factors are related to a dependent variable. Our approach with these situations is to perform a **multiple regression,** which means that several independent variables are included in the regression.

Fortunately, many concepts from Chapter 7 carry over to multiple regression with little or no modification. One difference, however, is that calculations for multiple regression are more tedious than for single variable regression. A computer is therefore an indispensable tool for multiple regression. The computer's ability to perform many calculations quickly is a primary reason that multiple regression is a predominant analytical tool today. In this chapter, we will not emphasize the calculations, but will instead focus on understanding multiple regression equations. We can only scratch the surface on this subject. Interested readers should refer to the statistics books cited in Chapter 1 for additional information.

> **Multiple regression** is regression with several independent variables rather than a single independent variable. Multiple regression is beneficial when the dependent variable is related to several factors. The resulting multiple regression provides a means of estimating the dependent variable based on values of the several independent variables. Many of the concepts from Chapter 7, such as coefficient of determination and standard error, can be applied to multiple regression.

Boston Fires

The example for this section is based on 1990 census tract information for Boston, Massachusetts combined with data on residential fires for 1989 and 1990 by census tract. A total of 147 Boston census tracts had at least one fire during these two years. A preliminary analysis with correlations showed that residential fires in the Boston census tracts were related to four key independent variables:

- POPULATION: Population in a census tract.
- BOARDED: Number of boarded-up housing units in a census tract.

- FAMTYPE: Number of single-parent households in a census tract with one or more persons under 18 years of age.
- DENSITY: Number of households in a housing unit with one or more persons per room.

As we shall see later, census tracts with large values for these independent variables tended to have more fires, and vice versa. On the basis of these tendencies, the four variables are good candidates for a multiple regression equation with fires as the dependent variable. The resulting multiple regression equation provides insight into reasons why fires are high in one area and low in another area.

Data for the four independent variables were collected as part of the 1990 census conducted by the Bureau of Census. Population is, of course, the number of persons residing in a census tract. The other three variables can be considered as measurements of the socio-economic conditions of census tracts. Census tracts with poor socio-economic conditions usually have more boarded-up buildings, more households headed by single parents, and more persons per room. The parallel between socio-economic conditions and fires is probably no surprise to fire fighters and fire department administrators. The advantage of multiple regression is that we can quantify the relationships between fires and socio-economic conditions rather than depending on personal experiences.

For this analysis the census tract data were combined with residential fire data fire each census tract. Merging the data was possible because the Boston Fire Department records the census tract on each fire incident report. Table 8-1 shows the averages and standard deviations for the variables in this analysis.

Exhibit 8-1 Averages and Standard Deviations for Regression Analysis		
Variable	Average	Standard Deviation
Residential Fires	13.1	9.0
POPULATION	1382.2	790.4
BOARDED	10.2	21.6
FAMTYPE	177.6	148.4
DENSITY	95.4	79.2
		Note Boston. Mass. Data

The average number of fires for the 147 census tracts was 13.1 with a standard deviation of 9.0. These statistics were derived directly from the Boston Fire Department data. The other averages and standard deviations were calculated from 1990 census data for the 147 census tracts.

Exhibit 8-2 shows the correlation matrix for these five variables. The

DENSITY variable has the highest correlation with residential fires (.74), followed by FAMTYPE (.61), BOARDED (.37), and POPULATION (-35). The correlations among the independent variables are also important to review. As we will discuss in the next section, we do not want a multiple regression with independent variables that are highly correlated with each other. Exhibit 8-2 shows that the largest correlation for the independent variables is .58 between DENSITY and FAMTYPE. Most of the other correlations are low. In fact, the correlation between POPULATION and BOARDED is -.06, which means that these two variables have virtually a random relationship.

Exhibit 8-2 Correlation Matrix-Boston Census Tracts-1990

Variable	FIRES	POPULATION	BOARDED	FAMTYPE	DENSITY
FIRES	1.00	.35	.37		
POPULATION	.35		-.06		
BOARDED	.37	-.06	1.00		
FAMTYPE	.61	.12	.43		
DENSITY	.74	.35	.23		

As previously indicated, computers are a necessity for multiple regression because of the complicated nature of the calculations. We will not attempt to show calculations as we did in Chapter 7, because our objective is on interpreting and applying a multiple regression equation. For the residential fires in Boston, the resulting regression equation was as follows:

$$\text{Fires} = 2.0 + .0017 \times \text{POPULATION} + .068 \times \text{BOARDED} + .013 \times \text{FAMTYPE} + .060 \times \text{DENSITY} \qquad (1)$$

We can estimate the number of fires in a census tract by knowing the values of the independent variables. For example, census tract 510 has 1,607 residents, 2 boarded-up units, 154 single-parent households, and 33 households with one or more persons per room. The estimated number of fires for this census tract is as follows:

$$\text{Fires} = 2.0 + .0017 \times 1607 + .068 \times 2 + .013 \times 154 + .060 \times 33 \qquad (2)$$
$$= 8.92$$

Census tract 510 experienced 10 fires so that the estimated number of fires is very close to actual experience.

Another interesting feature of the regression is that the coefficients for the independent variables in (1) are always positive. This means that the number of fires increases as these variables increase. According to the multiple regression results, increases in population, boarded-up units, household density, and single-parent households will result in increases in fires.

As with regressions in Chapter 7, we should he interested in how well the regression equation fits the actual data. One measure of fit is the **coefficient of determination,** which has exactly the same definition as in Chapter 7:

$$R^2 = \frac{SST - SSE}{SST} \tag{3}$$

where SST is the total sum of squares and SSE is the **sum of squares for errors.** *Formally,* SST and SSE are defined as follows:

$$SST = \sum (y_i - \bar{y})^2 \tag{4}$$

$$SSE = \sum (y_i - \hat{y}_i)^2 \tag{5}$$

where y_i are the actual values of the dependent variable, and $\bar{y}$ is the average of the dependent variable and $\hat{y}_i$ are the estimated values from (1). The coefficient of determination, R^2, is always between 0 and 1. A R^2 value close to 1 indicates a good fit while a value close to 0 indicates a poor fit.

R^2 increases whenever we introduce a new independent variable into the regression. However, we need to avoid the temptation to add independent variables just to increase this value. In practice, you will find that a few independent variables will increase R^2 considerably with additional variables adding very little to the R^2 Most computer programs have a procedure for selecting the most important independent variables and omitting variables that have minimal contribution to the regression equation. This procedure is called **stepwise regression** because it introduces independent variables one by one according to their importance until the inclusion of more variables does not significantly improve the equation. Most computer programs for multiple regression include a procedure for stepwise regression.

For the regression with Boston residential fires, SST is 11,868.7 and SSE is 4,333.8 so that R^2 is as follows:

$$R^2 = \frac{11,868.7 - 4,333.8}{11,868.7} \tag{6}$$
$$= .63$$

This value indicates a fairly good fit, although it is not as large as we would like for this type of analysis.

Another measure of interest is the **standard error,** which we defined in Chapter 7 as indicating how far, on average, the estimates from the regression deviate from the actual numbers. For multiple regression, the equation for the standard error is as follows:

$$\text{Standard Error} = \sqrt{\frac{SSE}{n - k - 1}} \tag{7}$$

where n is the number of points and k is the number of independent variables. For our regression, we calculate the standard error as follows:

$$\text{Standard Error} \equiv \sqrt{\frac{4333.8}{147 - 4 - 1}}$$

$$= 5.52$$

(8)

This standard error means that the estimates of fires will deviate, on average, by 5.52 from the actual number of fires for the census tracts.

In summary, the interesting feature of this regression is that four variables have been identified which can estimate the number of residential fires in a fairly accurate manner. The resulting regression equation could be employed by the Boston Fire Department for planning purposes. For example, other information in the city may be available on expected changes in the four independent variables. The department could therefore estimate its workload for these census tracts in future years.

> The **coefficient of determination**, R^2, determines how well a regression line fits the data points. R^2 is always between 0 and 1. Low values indicate a poor fit to the data while values close to 1 indicate a good fit to the data. The standard error indicates how far, on average, the estimates from the regression deviate from the actual values.

Collinearity Between Variables

A problem with multiple regression occurs when two independent variables are highly correlated. When this situation occurs, we need to select **one** of the variables to include in the regression. As an example, we present the data in Exhibit 8-3 from Prince William County, Virginia. The exhibit shows the total number of fires (residential and non-residential) in the county for 1981 through 1989. Two independent variables related to fires are shown in the last two columns. The column labeled "Residences" gives total number of residences (in thousands) and the last column labeled "Non-Residential Space" gives total square feet (in hundred thousands) of non-residential (retail, office, and industrial) space in the county. The data on residences and non-residential square footage are collected on an annual basis by the county.

Exhibit 8-4 shows the correlation matrix for these three variables. All correlations are high. The correlation between fires and residences is .90 and the correlation between fires and non-residential square footage is .91. The exhibit also shows a very high correlation of .98 between residences and non-residential square footage, which means that the pattern of annual increases is virtually the same for these two variables.

Exhibit 8-3 Fires in Prince William County, Virginia-1981-1989

Year	Total Fires	Residences	Non-residential Space
1981	3,313	47.91	106.98
1982	3,003	49.64	107.69
1983	2,938	50.93	108.51
1984	3,157	53.05	112.91
1985	3,631	54.81	125.10
1986	3,877	57.34	144.13
1987	3,761	60.57	162.57
1988	4,256	65.96	182.77
1989	4,156	68.65	206.03

Exhibit 8-4 Correlation Matrix, Prince William County, Virginia

Variable	Fires	Residences	Non-residential Space
Fires	1.00	.90	.91
Residences		1.00	.98
Non-residential Space	.91	.98	1.00

The temptation is to include both variables in a regression equation with fires as the dependent variable. However, we will now show that a regression with only one of the variables gives virtually the same fit to the data as a regression with both variables. Exhibit 8-5 shows the regression equations and coefficients of determination (R^2) for a regression with residences (TOTRES) as the only independent variable, non-residential space (TOTFT) as the only variable, and both variables in a regression.

Exhibit 8-5 Regression Equations for Fires in Prince William County

Variables in Regression	Regression Equation	R^2
TOTRES	Fires = 60.8 x TOTRES + 129.5	.816
TOTFT	Fires = 12.1 x TOTFT + IJ76.9	.829
Both	Fires = 13.9 x TOTRES + 9.4 x TOTFT + 1,469.9	.830

The three regression equations differ considerably because they include different independent variables. However, the coefficients of determination allow us to make comparisons. We have a R^2 value of .816 for the regression with only TOTRES and .829 for the regression with only TOTFT. With both variables in the regression, the coefficient of determination increases very minutely to .830. Consequently, the regression with only TOTFT gives almost exactly the same fit as including both variables. The inclusion of TOTRES with TOTFT contributes virtually nothing to the fit of the regression equation to the data.

The high correlation between TOTRES and TOTFT is the reason that we have such a small increase in the R^2 value when both variables are included. In summary, the best approach in this example is to select the regression equation with TOTFT to estimate fires rather than using the regression equation with both variables.

The term **multicollinearity** is used in most textbooks on statistics to refer to the situation in which high correlations exist between independent variables. In addition to problems with selection of variables, multicollinearity can result in instability of the estimated coefficients. The instability means that estimated coefficients may vary considerably from one sample to the next.

> Multicollinearity is the term used to indicate that two or more independent variables are highly correlated with each other. When multicollinearity occurs, we usually want to select one of the correlated variables to include in the multiple regression. The other correlated variables will not improve the multiple regression and may, in fact, result in unstable coefficients in the equation

Regression with Dummy Variables

The examples in this chapter have been based on continuous variables. The regression for residential fires in Boston included population, boarded-up housing units, single-parent families, and high-density housing units. In previous chapters, however, we have noted that categorical variables are important to fire departments because the 901 codes serve as the basis for completing reports on fires. In this section, we will present an approach that includes categorical variables in a regression.

The example we present is based on data collected during 1990 on Emergency Medical Services (EMS) calls in Prince William County, Virginia. We have selected travel time to EMS calls as our dependent variable. Our aim is to determine the effect of three categorical variables on travel time:

- Area of origin of EMS calls
- Whether delays occurred enroute to calls
- Type of call (Advanced Life support (ALS) or Basic Life Support (BLS) call).

Chapter 8

For this analysis, three areas of the county were selected, which we will designate as Area A, Area B, and Area C: Each EMS report indicates the area of the call. In addition, the responding paramedic indicates whether delays were encountered while enroute to the scene. Delays may occur for several reasons, including traffic, weather, and incomplete address information.

Exhibit 8-6 shows average travel times for each variable. The overall average travel time for these 1,620 EMS calls was 6.71 minutes. As seen in the exhibit, the average travel times in the areas differ considerably with a low average time of 5.52 minutes in Area A followed by 7.99 minutes in Area C and 9.33 minutes in Area B. Area B and Area C had longer average travel times primarily because these areas arc larger than Area A. The exhibit also shows only a small difference between travel times to ALS and BLS calls. Travel times to BLS calls averaged 6.52 minutes compared to 7.11 minutes for ALS calls. Finally, the average trawl time to delayed calls was about one minute longer than calls not encountering delays (6.62 minutes compared to 7.69 minutes).

Exhibit 8-6 Average Travel Times-Prince William County-1990

Variable	Average Travel Time (Minutes	Number
Area A	5.52	955
Area B	9.33	218
Area C	7.99	447
ALS Calls	7.11	525
BLS Calls	6.52	1,095
No Delays	6.62	1,473
Delays	7.69	147
Overall	6.71	1,620

While Exhibit 8-6 provides useful information about travel times, it does not give any information on combinations of the variables. We do not know, for example, what travel time to expect in Area A for delayed ALS calls. We can employ regression analysis to provide estimates of travel times for the various combinations.

Since regression analysis requires numerical values, we need to assign numbers to these categorical variables. A convenient approach is to define new variables for each EMS record with appropriate numerical values depending on the type of call, area, and whether delays were encountered. For example, we define a variable called CALLTYPE to indicate the type of call. If an EMS record is for a BLS call, we assign a value of 0 to CALL-TYPE. If it is an ALS call, we assign a value of 1 to CALLTYPE. In a similar manner, we define a variable called DELAYS for each EMS incident. If delays were not encountered, we assign a value of 0 to DELAYS, and if delays were encountered, we assign a value of 1 to DELAYS.

Because we have three areas, we need a slightly different coding approach for them. We define two new variables called AREA1 and AREA2. If an EMS record is for a call in Area 1, we assign a value of 1 to AREA1 and a value of 0 to AREA2. Similarly, if the EMS call is from Area 2 we assign a value of 0 to AREA1 and a value of 1 to AREA2. Finally, if a call is from Area 3, we assign a value of 0 to both AREA1 and AREA2. Note that we would never assign a value of 1 to both AREA1 and AREA2.

In summary, we are defining four new variables in the following manner:

Variable	Definition
CALLTYPE	0 for ALS calls 1 for BLS calls
DELAYS	0 for calls not delayed 1 for delayed calls
AREA1	0 if call is not from Area 1 1 if call is from Area 1
AREA2	0 if call is not from Area 2 1 if call is from Area 2

The terms **dummy variable or indicator variable** are sometimes used to designate variables defined in this manner as either 0 or 1. A dummy variable indicates the presence or absence of a characteristic. The variable DELAYS indicates that a call was either delayed or not delayed; similarly, the variable CALLTYPE indicates that a call is either a BLS call or not a BLS call (that is, it is an ALS call). By assigning the values of 0 or 1, dummy variables transform categorical variables into meaningful numerical variables amenable to statistical analysis. In particular, we can perform a regression analysis with dummy variables, and the interpretation of the regression will be related to the definitions of these dummy variables.

For the regression, the dependent variable is travel time, and the independent variables are the four dummy variables. The resulting regression equation for the data from Prince William County is as follows:

$$\text{Travel Time} = 5.27 + 1.38 \times \text{DELAYS} + .36 \times \text{CALLTYPE} + 3.83 \times \text{AREA1} + 2.48 \times \text{AREA2} \tag{9}$$

Chapter 8

The regression equation indicates that we start with a base travel time of 5.27 minutes and then increase the travel time based on delays, type of call, and area of origin. For example, suppose that we want to estimate the travel time for a delayed ALS call from Area 2. The values of the dummy variables for this example would be as follows:

DELAYS	1
CALLTYPE	0
AREA1	0
AREA2	1

We assign a value of 1 to DELAYS because we are assuming a delayed call. A value of 0 is assigned to CALLTYPE because we are assuming an ALS call. The variable AREA2 is assigned a 1 because we are assuming a call from AREA 2, which means that a value of 0 is assigned to the variable AREA 1

Inserting these values into the regression equation gives the following result:

$$\text{Travel Time} = 5.27 + 1.38 \times 1 + .36 \times 0 + 3.83 \times 0 + 2.48 \times 1$$
$$= 9.13 \text{ minutes}$$

Exhibit 8-7 shows a systematic way of estimating travel times which takes advantage of the fact that we are working with dummy variables. With the exhibit, we start with a travel time of 5.27 minutes and then make additions depending on the values assigned to the dummy variables.

Exhibit 8-7 Travel Times Estimates From Regression

1. Start with a travel time of 5.27 minutes.

2. If the call is delayed, add 1.38 minutes.
   If the call is not delayed, add nothing.

3. If the call is a BLS call, add .36 minutes.
   If the call is an ALS call, add nothing.

4. If the call is from Area 1, add 3.83 minutes.
   If the call is from Area 2, add 2.48 minutes.
   If the call is from Area 3, add nothing.

An interesting result can be derived from determining the range of travel times. The smallest travel time will occur for non-delayed BLS calls from Area 3. Calls with these characteristics are estimated to have an average travel time of 5.27 (since we are adding nothing to the base average). On the other hand, delayed BLS calls from Area 1 will have the highest average travel time, which is estimated to he 10.83 minutes.

This example shows that dummy variables enable us to include categorical variables in a regression. We always have one fewer dummy variable than the number of levels in a categorical variable. We needed only one dummy variable to indicate delayed/non-delayed calls since there are only two categories. We needed two dummy variables for areas since we had three areas. You should he aware that a regression with all dummy variables is equivalent to another area of statistical analysis called **analysis of variance.** Details on analysis of variance are beyond the scope of this handbook. The key point is that analysis of variance and dummy variable regression are equivalent statistical procedures.

Finally, it should be mentioned that we can combine continuous and dummy variables in a regression analysis. That is, it is not necessary to have either ail continuous variables or all dummy variables in a multiple regression. The combination of variables in sometimes called **analysis of covariance.**

> Dummy variables or indicator variables are variables that have values of either 0 or 1. A dummy variable indicates the presence or absence of a characteristic. Dummy variables provide a means of converting categorical variables to numerical variables amenable to statistical analysis. The number of dummy variables for a categorical variable is always one less than the number of levels of the category variable.

Summary

Multiple regression means that several independent variables are included in the regression analysis. The advantage of multiple regression is that it allows us to determine the impact of these independent variables on the dependent variable. The independent variables may he continuous or categorical. For categorical variables, we must develop dummy variables, which indicate the presence or absence of a characteristic, for the regression. As with single variable regression in Chapter 7, we can calculate a coefficient of determination and a standard error to determine how well our regression fits the actual data. The equations for these are very similar to their counterparts for single variable regression.

Chapter 8

PROBLEMS

1. The regression equation for the Boston census tracts was follows:

$$\text{Fires} = 2.0 + .0017 \times \text{POPULATION} + .068 \times \text{BOARDED} \quad (1)$$
$$+ .013 \times \text{FAMTYPE} + .060 \times \text{DENSITY}$$

Estimate the number of fires for each of the following census tracts and compare your result to the actual number of fires shown in the last column.

Census Tract	Population	Boarded Units	Single Parent families	Housing Density	Actual Fires
709	1386	19	184	100	15
812	980	148	396	180	30
902	664	59	261	104	30
907	1311	3	139	66	10

2. The multiple regression from Prince William County), for travel times to EMS calls is as follows:

$$\text{Travel Time} = 5.27 + 1.38 \times \text{DELAYS} + .36 \times \text{CALLTYPE} \quad (9)$$
$$+ 3.83 \times \text{AREA1} + 2.48 \times \text{AREA2}$$

Estimate the travel times for the following types of calls:

a. A non-delayed BLS call from Area 1

b. A delayed ALS call from Area 2.

c. A non-delayed ALS call from Area 3.

d. A delayed BLS call from Area 3.

QUEUEING ANALYSIS

Applications of Queueing Theory

Waiting lines have become an everyday part of our lives. Think for a moment about your activities during the past few days and how often you have waited in line for some type of service-at a post office, grocery store, theater, hank, airport, or cafeteria. The common feature of these diverse situations is that people arrive for a service that is unavailable because the providers are busy. Waiting lines present a dilemma for managers responsible for the delivery of service. For example, if a bank manager does not have enough tellers, the waiting line of customers may become quite long and they may eventually switch to another bank. On the other hand, if a manager has many tellers on duty, tellers will frequently be idle and bank costs will escalate.

Within many fire departments, we find a similar dilemma in determining how many EMS units we need to adequately handle requests from citizens for emergency medical services. A "waiting line" develops when all EMS units are busy and more citizens call with medical problems. The waiting line is actually in the communications center where the calls must be held until EMS units become available. If there are not enough units in the field, citizens will occasionally have to wait because everyone is busy. On the other hand, if the department fields a large number of EMS units, then costs will increase and units may not be very busy.

Queueing theory provides methods to analyze whether waiting lines will occur and what the consequences of waiting lines will be. It can be applied to diverse situations including the determination of the number of tellers for a bank, the number of ticket agents at an airport, and the number of ambulances for fire departments. In this chapter, we will apply a queueing model to determine the number of ambulances needed.

The following quote from an issue of the **jems** magazine reiterates the points we have been making and serves as an introduction to the topics covered in this chapter."

> No one likes to wait in a line, and none of the systems
> with which we work would permit a lengthy wait to occur
> for a patient while the dispatcher searches for an ambu-
> lance to send on the critical call. On the other hand, no
> system can place an ambulance on every corner and
> call in a backup crew when each call is dispatched.
> Some where between having too many or too few ambu-

13. Barton. George K. "The Wait for an Ambulance." jems. December, 1986

Chapter 9

lances a viable solution: queueing theory. This simple mathematical model or set of formulas will enable you to determine the number of ambulances needed, by hour of day and day of week, to meet calls for service in an efficient manner, and will provide objective information to modify previously committed resources.

As indicated in this quotation, queueing theory provides an analytical procedure for determining how many ambulances are needed by hour of day and day of week. In order to make these calculations, however, you must know how many EMS calls you expect and the average amount of time ambulance crews will take on calls. We can make these determinations using the techniques previous described in Chapter 4 of this handbook.

> Queueing theory is the study of waiting lines and the consequences of these lines. It has been applied in a variety of situations where waiting lines occur. It is an excellent approach for fire departments in determining how many EMS units they need to field to handle calls from citizens for emergency medical services.

In summary, queueing theory can provide assistance in managing and operating the emergency medical services for a fire department. Potential applications include:

- Estimation of how busy EMS units will be based on an expected workload of EMS calls.
- Estimation of the probability that a citizen's call for medical services will have to wait because all EMS units are busy.
- Estimation of the average number of citizens waiting for medical service.
- Estimation of the average waiting time for these citizens.
- Estimation of the number of EMS units needed to satisfy objectives established by a fire department.

Examples of Results from Queueing Theory

Unit Utilization

As a starting point, suppose you have determined that during a particular time period (e.g., on Saturday evenings from 8 p.m. to midnight), citizen calls for EMS service arrive into the communications center at an average rate of two calls per hour. Assume that 40 minutes is the mean time for a call. The time required for a call starts with the dispatch of an EMS unit to the scene and ends when the unit is available for another call.

Suppose further that two EMS units are available in the field to respond to citizen calls. As a starting point for our analysis, we can calculate how busy these two units will be. They provide 120 minutes of unit time each hour (2 units times 60 minutes). The calls require 80 minutes each hour (2 calls times 40 minutes each). Thus, each unit will be busy about 66.7 percent of the time (80 minutes of work divided by 120 minutes of unit time). This percentage is called "unit utilization" since it measures the percent of time each unit will be busy during a given time period.

If three units are fielded, unit utilization drops to 44.4 percent, as seen by the following calculation:

$$\text{Unit Utilization} = \frac{2 \times 40}{3 \times 60} = 44.4\%$$

In this equation, the numerator gives the amount of work we expect and the denominator gives the amount of available unit time. You can use this equation to verify that with four units, the unit utilization drops to 33.3 percent, and with five units, it is 26.7 percent. Exhibit 9-1 is a graph of unit utilization for this example. Of course, unit utilization can never equal zero, but it continually decreases as we add more units in the field.

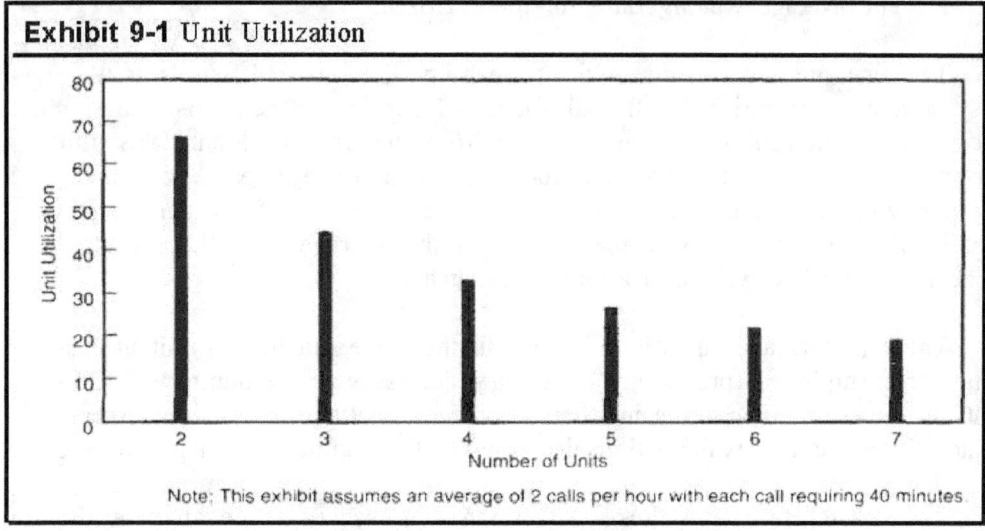

Exhibit 9-1 Unit Utilization

Note: This exhibit assumes an average of 2 calls per hour with each call requiring 40 minutes.

We can state a general formula for unit utilization as follows. Assume that we expect citizen calls for EMS service each hour and that calls average t minutes from time of dispatch to time of completion. If we field n units, the unit utilization is expressed by:

$$\text{Unit Utilization} = \frac{ct}{60n}$$

In practice, we can expect considerable variation from what has just been described. Some evenings will be busier than others and some calls will require less time than average, while other calls will take considerably longer. The point is that we will not have exactly the same unit utilization every day; it will fluctuate depending on the actual workload experienced. When we calculate unit utilization, we are looking at what happens "on average." While we expect fluctuations, we are primarily concerned with overall average performance of our system.

Other Queueing Calculations

Unit utilization is an example **of a system performance measure.** It measures how the EMS system will perform under expected workload conditions. We are not evaluating individual units and we arc not trying to determine the unit utilization of individual units. Instead, we are considering the system as a whole and making estimates on what we expect to occur.

By applying queueing theory, we can estimate several other system performance measures. Three of the most common are:

- The probability that a person calling for service will have to wait.
- The average number of citizens waiting for service.
- The average waiting time for these citizens.

The first measure estimates the probability that all EMS units will be busy and another citizen will call for EMS service. When this situation occurs, the communications center must hold the call until an EMS unit becomes available. The second performance measure estimates how many citizens will be waiting, on average, for the dispatch of an ambulance. Finally, the third performance measure gives the average time these waiting citizens will have to wait until a unit is dispatched.

These performance measures behave in the same manner as unit utilization. For example, the probability of a delay decreases as the number of EMS units increases. Similarly, the number of citizens waiting for service decreases and the amount of waiting time decreases as the number of units increases.

The reason we have these performance measures is because of the inevitable variation in workload. If two units were fielded, no waiting lines would occur if we were assured that we would always have exactly two calls per hour requiring 40 minutes each. In reality, however, workload varies, and waiting lines develop because of these variations. Since queueing models assume considerable variability, they provide excellent estimates on the length of waiting lines and other performance measures.

The formulas for the probability. of delay and the length of a waiting line arc complicated. Interested readers arc invited to study the next section where we provide explanations for both calculations. Appendices B and C have been included in this handbook to ease the calculation burden. They provide a quick way to obtain the probability of delay and length of a waiting line without having to perform many calculations. The tradeoff is that the results may not be quite as precise as the actual formulas because of the rounding of numbers we have to do prior to access into the tables. For most applications, the errors will be small. However, we strongly suggest that you try to use the actual formulas presented later in this chapter if you want to be precise in your analysis.

The key to the tables in Appendices B and C is the following calculation with the average number of calls and average time per call.

$$\text{Table Key} = \frac{ct}{60}$$

where c is the number of calls per hour and t is the average time per call. Note that division by 60 converts the average time to hours for convenience with the tables. The Table Key is actually the amount of work we expect each hour.

In our example, c is 2 calls per hour and t is 40 minutes. The Table Key is then 1.33 (2 x 40 divided by 60). We round to 1.3 since the keys in the appendices are to one decimal place. Now turn to Appendix B to find the probability that a call will be delayed. The number of units is shown across the top of the table and the Table Key is shown down the left column. Move down the left column to the line where 1.3 appears as the key. You can now read across this row to obtain the probability of a delay. With two units, the probability is 51.2 percent; with three units, the probability is 17.0 percent, etc.

To determine the average number of citizens waiting for an EMS unit, we use the table in Appendix C in the same manner. The number of units is displayed across the top and the Table Key is down the left column. We go to the line where 1.3 appears as the key and obtain the average number of waiting calls by moving across the row. With two units, we estimate .95 citizens waiting; with three units, only .13 citizens waiting; and with four units, only .02 citizens waiting.

Finally, we want to determine the amount of time that waiting citizens will have to wait before a dispatch occurs. Queueing theory informs us that to determine the average waiting time, we simply divide the entry from Appendix C by the average number of calls per hour:

$$\text{Waiting Time} = \frac{\text{Appendix C Entry} \times 60}{c}$$

Note that we have multiplied by 60 so that our final answer will be in minutes rather than hours. For example, with two units, we have determined that .95 citizens will be waiting. Since we have 2 calls per hour, the average waiting time will be 28.5 minutes (.95 times 60 divided by 2).

The top of Exhibit 9-2 summarizes the results of our calculations. The averages of 2 calls per hour and 40 minutes per call are based on data from Prince William County, Virginia, for 1990. By way of comparison, the bottom portion of the exhibit shows how performance measures change when the average number of calls increases to 2.5 calls per hour. They show dramatic increases. With three units, for example, unit utilization changes from 44.4 percent to 55.6 percent, and the probability of a delay goes from 17.0 to 30.0. On the basis of this exhibit, the county might decide to increase the number of units because of the changes in performance measures caused by the increased workload.

Exhibit 9-2 Summary of Performance Measures

Assuming 2.0 Calls Per Hour and 40 Minutes Per Call

	Number of Units			
Performance Measure	2	3	4	5
Unit Utilization	66.7 %	44.4 %	33.3 %	26.7 %
Probability of a Delay	51.2	17.0	4.8	1.1
Average Number of Waiting Calls	0.95	0.13	0.02	0.005
Average Waiting Time for These Calls (Minutes)	28.5	3.9	0.6	0.2

Assuming 2.5 Calls Per Hour and 40 Minutes Per Call

	Number of Units			
Performance Measure	2	3	4	5
Unit Utilization	83.3 %	55.6 %	41.7 %	33.3 %
Probability of a Delay	75.8	30.0	10.3	3.0
Average Number of Waiting Calls	3.79	0.37	0.07	0.015
Average Waiting Time for These Calls (Minutes)	90.9	9.0	1.8	0.4

Note. For the bottom portion of this exhibit, exact calculations with the queueing theory formulas were used

---

Determining the Number of EMS Units

The prior analysis presented queueing theory as a descriptive tool. That is, we assumed a fixed number of EMS unit in the field and estimated the performance measures. We can also employ queueing theory in a **prescriptive** manner to determine how many units will be required to achieve predetermined performance measures. That is, we establish objectives for performance and determine how many units will be needed to achieve the objectives. This approach is similar to a management-by-objectives approach in which objectives determine performance standards.

Suppose, for example, that your objectives are to field enough EMS units to insure that:

- Units are busy on EMS calls no more than 40 percent of their shift.
- The probability of a delay is 5 percent or less.

What we want to determine is the number of units needed to obtain these performance standards.

For purposes of illustration, assume that we expect 3.2 calls per hour averaging 37 minutes each. The first objective is on unit utilization. We can use our equation for unit utilization and express it in terms of units needed:

$$\text{Units Needed} = \frac{ct}{60 \times \text{Unit Utilization}}$$

Since our objective is 40 percent for unit utilization, we obtain the number of units needed as follows:

$$\text{Unit Needed} = \frac{3.2 \times 37}{60 \times .40} = 4.93$$

We round this result to 5 units since we cannot have a fractional unit. This calculation means that five units will achieve our objective of no more than 40 percent of their time devoted to EMS calls. The result assumes, of course, that the jurisdiction will continue to average 3.2 EMS calls per hour and 37 minutes per call.

To estimate the number of units needed for the second objective, we reference the table in Appendix A. The objective is that the probability of a delay will be 5 percent or less. To access the table, we need the Table Key, which calculates to 1.97 (3.2 times 37 divided by 60). We round this result to 2.0 in order to access the table. We now look at the row for 2.0 in Appendix A, which is shown as Exhibit 9-3 on the following page.

This table indicates that six units will be needed to achieve our objective that the probability of delay will not exceed 5 percent. With three units or four units, the probability of delay is higher than we are willing to tolerate at 44.4 percent and 17.4 percent, respectively. With five units, the probability of delay will be only 6.0 percent, which is close to our objective but does not achieve it. We therefore select six units to he assured of satisfying our objective.

Exhibit 9-3 Probability of Delay

Table Key = 2.0 (3.2 calls averaging 37 minutes each)

Number of Units	Probability of Delay
3	44.4
4	17.4
5	6.0
6	1.8
7	.01

We have now determined how many units are needed for each objective, but we still have one remaining step. We need five units to achieve the first objective, and we need six units to achieve the second objective. To achieve **both** objectives, we select the **maximum** of these two numbers. In other words, we select six units as the final answer. The reason is that six units will satisfy both objectives. It satisfies the second objective because of our manner of selection from the table, and it satisfies the first objective because adding more units decreases unit utilization.

As a final note, it should be mentioned that there are many other queueing models from which to choose. We have selected this particular model because it is one of the most frequently applied and has proven beneficial in many queueing problems similar to what fire departments encounter.

Queueing models can he expanded to cover other situations. For example, we can include call priorities to take into account that some EMS calls are more serious than others. If citizen calls are waiting in the communications center, the more serious calls, such as heart attacks, have higher priority than minor calls, such as bruises. Queueing models exist that include priority systems. They can be applied in the same manner as our basic model. In addition, other queueing models exist that take geography into consideration. These models are more complicated because they require the user to describe the geography of your jurisdiction. However, they may be beneficial to you because they may reflect more accurately the problems of locating EMS units.

Queueing Calculations*

In the literature on queueing theory, the system just described is called a multiple-server queueing model in which call arrivals follow a Poisson distribution and service times follow an exponential distribution. In this section, we will describe the Poisson and exponential distributions and we will give formulas for the queueing calculations needed to derive Appendices A and B.

The Poisson distribution is a probability distribution which has been applied in many diverse situations. To see its application to EMS calls, we will first develop a frequency distribution on the number of calls arriving into a communications center. The frequency distribution will then be approximated by the Poisson distribution.

During some hours, no calls will come into the communications center. In other hours, we will have only one EMS call, and other hours will have more than one EMS call. Over a large number of hours, we can build a frequency table showing the number of hours with zero calls, the number with only one call, the number with two calls, etc.

As an example, Exhibit 9-4 gives information on the number of calls per hour in Prince William County, Virginia, for the hours from 3 p.m. to 7 p.m., January 1-May 31, 1990 (a total of 152 days). Since we are looking at four hours during each of these days, we have 608 data points (4 times 152). The exhibit shows that there were 54 hours during which no citizens called for EMS service. For 105 hours, exactly one EMS call came into the communications center; for 123 hours, exactly two EMS calls came into the communications center, and so on. The third column of the exhibit converts the frequencies into percentages. This column shows, for example, that during 14.0 percent of the hours, 4 EMS calls arrived into the communications center.

EMS Calls	Number of Hours	Percent	Poisson
0	54	8.9	5.2
1	105	17.3	15.3
2	123	20.2	22.7
3	109	17.9	22.4
4	65	14.0	16.6
5	51	8.4	9.8
6	37	6.1	4.9
>6	44	7.2	3.2
Total	608	100.0	100.0

Note: Calls are from Januay - May, 1990-3 p.m. to 7 p.m.

From Exhibit 9-4, we can calculate that the average number of EMS calls per hour is 2.97. We can then develop a Poisson distribution of the expected percentage of calls per hour. The right column of Exhibit 9-4

Chapter 9

shows expected frequencies according to a Poisson distribution (we will give the calculations for this column in the following paragraphs). This column can be compared to the actual percentage from our sample. While the expected percentages are not exactly the same, the differences are not large. For example, the Poisson distribution shows 15.3 percent of the hours with exactly one EMS call per hour, as compared to actual experience of 17.3 percent.

The general equation for a Poisson distribution is as follows:

$$P(x=k) = \frac{e^{-c}c^k}{k!}$$

where c is the average number of calls per hour, and k is an integer value starting with zero.

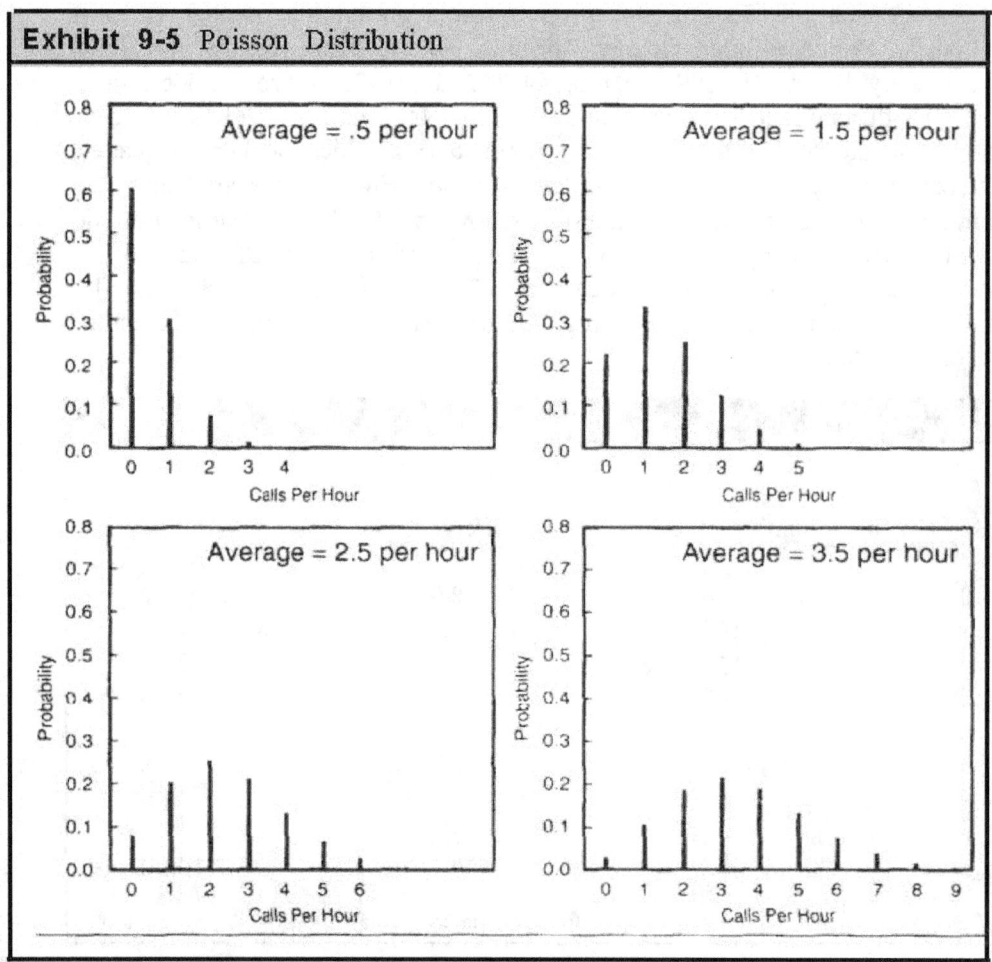

Exhibit 9-5 Poisson Distribution

In our example, we have an average of 2.97 calls per hour. To determine the probability of four calls per hour, we compute as seen on the following page.

$$P(x=4) = \frac{e^{-2.97} \times 2.97^4}{4!} = \frac{77.81 \times .0513}{24} = .166$$

Exhibit 9-5 shows several theoretical Poisson distributions. The shape of the distribution depends on the average. When the average is less than 1.0, the Poisson distribution continually decreases; if the average is greater than 1.0 the distribution increases to the integer portion of the average and then decreases.

The results on queueing theory presented in this chapter are correct provided the arrival of EMS calls into the Communications Center approximately follow a Poisson distribution. Hoaglin, Mosteller, and Tukey (1985) provide a relatively simple graphical technique for determining whether the Poisson distribution is a good selection. In their approach, k represents the number of possible EMS calls in an hour (0, 1, 2, etc.), and n_k represents the observed number of hours with k calls. If calls follow a Poisson distribution, you should obtain a straight line by plotting k against the quantity log *(k! nk! N)*, where N is the total number of hours (608 in our example). Exhibit 9-6 shows the plot obtained through this procedure. You will note that the dots approximate a straight line, which indicates a relatively good fit between our actual data and what would be expected with a Poisson distribution.

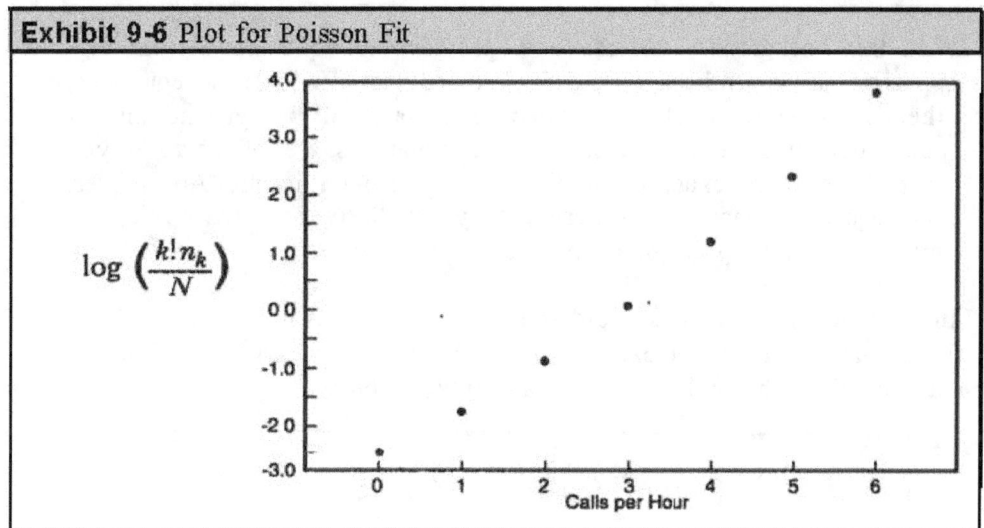

Exhibit 9-6 Plot for Poisson Fit

As previously indicated, the queueing model also assumes that the service times follow an exponential distribution. The exponential distribution is given by:

$$f(x) = ue^{-ux}$$

where x represents a particular service time and u is the **service rate.** Suppose, for example, that the average service time for EMS calls is 28.5 minutes, or .475 hours. The service rate is the inverse of the service time; that is, it is 1 divided by the service time. The service rate is therefore 2.11

Exhibit 9-7 shows the exponential distribution assuming a service rate of 2.11. The key feature of the exhibit is that the curve continually decreases. The decrease means that there is a low probability of long service times.

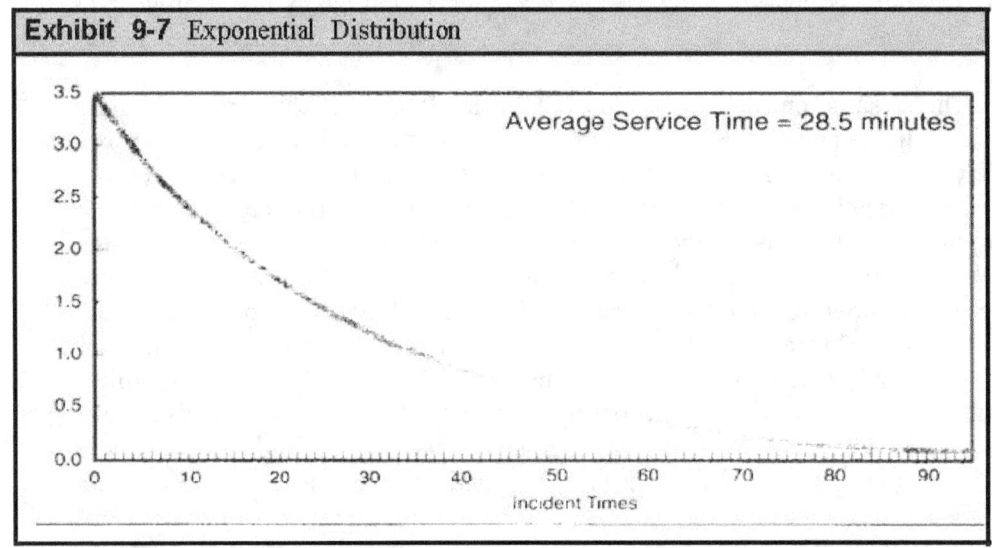

Exhibit 9-7 Exponential Distribution

With this background, we can now provide the equations for the figures appearing in Appendices A and B. We will not derive these equations, since the derivation assumes a considerable amount of background knowledge about queueing theory. However, the equations are of value if you want to perform more exact calculations than provided in the Appendices. While the equations appear complicated, they are fairly easy to develop on most microcomputer spreadsheet program.

In the following equations, we define $r = c/nu$. We assume that $r < 1$, so that the arrival rate does not exceed the maximum service rate. We first calculate P_0, which is the probability that all units are busy:

$$P_0 = 1 \bigg/ \sum \frac{\left(\frac{c}{u}\right)^k}{k!} + \frac{\left(\frac{c}{u}\right)^n}{n!} \frac{1}{(1-r)}$$

Then the probability of a delay is expressed as:

$$\text{Probability of Delay} = \frac{P_0 \left(\frac{c}{u}\right)^n}{n! \, (1-r)}$$

Finally, the equations for the queue length, L_q, and for waiting time in the queue, W_q, are given by:

$$L_q = \frac{P_0 \left(\frac{c}{u}\right)^n r}{n! \, (1 - r)^2}$$

$$W_q = \frac{L_q}{c}$$

Summary

Queueing theory is the analysis of waiting lines and the consequences of these waiting lines. It can be a powerful management tool to address issues of unit utilization, waiting time, call priorities, and other important considerations in your EMS delivery system. In this chapter, we have applied queueing theory to the problem of determining the number of EMS units needed in a fire department. For this determination, we must first know how many EMS calls we expect and what the average time per EMS call will be. We obtain these averages from prior experience using the techniques described in Chapter 4 of this handbook. It is also necessary to establish objectives for the EMS delivery system. We can say, for example, that we want to have enough EMS units so that they devote no more than 50 percent of their time to EMS calls. Other objectives could be established on the probability of a call delay, the average number of waiting calls, and the average that calls will have to wait. We can then apply a queueing model to determine how many units will be needed to achieve these objectives.

Chapter 9
PROBLEMS

1. Suppose that your department averages 3.5 EMS calls per hour with an average of 45 minutes for each call.

   a. Calculate unit utilization assuming you have 3, 4, 5 and 6 units fielded.

   b. How much does unit utilization change from 4 units to 5 units? from 5 units to 6 units?

2. For the average of 3.5 calls per hour and 45 minutes per call, use Appendices A and B to answer the following questions assuming 3, 4, 5, and 6 units.

   a. What is the probability, of delay?

   b. What will be the average number of citizens waiting for service?

   c. What is the average waiting time of citizens whose call has been delayed?

3. Suppose that EMS calls have been increasing 6 percent per year and that next year you also expect that calls will average 50 minutes per call, rather than 45 minutes per call.

   a. How many calls per hour will you have based on a 6 percent increase assuming a current average of 3.5 calls per hour.

   b. Determine unit utilization, probability of delay, and average number of people waiting assuming 4 units in the field.

   c. Compare these results to your answers in question 2 for 4 units.

4. How would you determine the actual value of your performance measures in your department today

5. What process would you use to set objectives in your jurisdiction on performance measures?

6. Suppose you expect 4 EMS calls per hour next month with an average of 38 minutes per call. Suppose further that your jurisdiction has set the following objectives:

   - Unit utilization not more than 60 percent
   - Probability of a delay not more than 3 percent

- Average number of citizens waiting not to exceed .25

a. Determine how many units will be needed to satisfy each of these objectives.

b. How many units will be needed to satisfy all three objectives?

7. Suppose the jurisdiction decides to add an objective that the waiting time should not exceed 15 minutes.

a. How many units will be needed to satisfy this objective?

h. Does this change the total number of units needed to meet all four objectives?

8. In Prince William County, Virginia, the distribution of EMS calls coming into the communications center between 7 a.m.-11 a.m., January-May 1990, was as follows:

EMS Calls	Number of Hours	Percent
0	158	26.0
1	192	3 1.6
2	144	23.7
3	68	11.2
4	34	5.6
5	10	1.6
6	2	0.3
Total	608	100.0

From this distribution, we can calculate an average of 1.46 calls per hour.

a. with the average of 1.46 calls, determine the expected percent of EMS calls under a Poisson distribution.

b. Develop a plot for Poisson tit similar to Exhibit 9-6 to determine whether the Poisson distribution gives a good approximation to the experienced distribution.

c. What is your conclusion?

9. For an exponential distribution, the cumulative distribution is given by the following relationship:

$$F(x) = 1 - e^{-xu}$$

where u is the service rate in our application.

Chapter 9: Problems

(continued from question 9)

Suppose that the average service time is 28.5 minutes, which means that the service rate is 2.11.

a. Develop a graph of the cumulative distribution assuming this service rate of 2.11.

b. From the graph, estimate the 25th percentile, median, and 75th percentile.

CRITICAL VALUES FOR THE CHI-SQUARED DISTRIBUTION

Degrees of Freedom ν	Critical Value .050	Degrees of Freedom ν	Critical Value .050
1	3.841	79	100.7
2	5.991	80	101.9
3	7.815	81	103.0
4	9.488	82	104.1
5	11.07	83	105.3
6	12.59	84	106.4
7	14.07	85	107.5
8	15.51	86	108.6
9	16.92	87	109.8
10	18.31	88	110.9
11	19.68	89	112.0
12	21.03	90	113.1
13	22.36	91	114.3
14	23.68	92	115.4
15	25.00	93	116.5
16	26.30	94	117.6
17	27.59	95	118.8
18	28.87	96	119.9
19	30.14	97	121.0
20	31.41	98	122.1
21	33.67	99	123.2
22	33.92	100	124.3
23	35.17		
24	36.42		
25	37.65		
26	38.88		
27	40.11		
28	41.34		
29	42.56		
30	43.77		
31	44.98		
32	46.19		
33	47.40		
34	48.60		
35	49.80		
36	51.00		
37	52.19		
38	53.38		
39	54.57		
40	55.76		
41	56.94		
42	58.12		
43	59.30		
44	60.48		
45	61.66		
46	62.83		
47	64.00		
48	65.17		
49	66.34		
50	67.50		
75	96.22		
76	97.35		
77	98.48		
78	99.62		

$Pr = P(X' \geq c)$

c = critical value

Appendix A

PROBABILITY OF DELAY

C/S		Number of Units 3	4	5	6	7
1	33.3	9.1	2.0	0.4	0.1	0.0
1.1	19.0	11.5	2.8	0.6	0.1	0.0
1.2	45.0	14.1	3.7	0.8	0.2	0.0
1.3	i1.?	17.0	4.8	1.1	0.2	0.0
1.4	57.6	20.2	6.0	1.5	0.3	0.1
1.5	64.3	23.7	7.5	2.0	0.5	0.1
1.6	71.7	27.4	9.1	2.6	0.6	0.1
1.7	78.1	31.3	10.9	3.3	0.9	0.2
1.8	85.3	35.5	12.9	4.0	1.1	0.3
1.9	92.6	39.9	15.0	4.9	1.4	0.4
2		44.4	17.4	6.0	1.8	0.5
2.1		49.2	19.9	7.1	2.2	0.6
2.2		54.2	22.7	8.4	2.7	0.8
2.3		59.4	25.6	9.8	3.3	1.0
2.4		64.7	28.7	11.4	4.0	1.3
2.5		70.2	32.0	13.0	4.7	1.5
2.6		75.9	35.4	14.9	5.6	1.9
2.7		81.7	39.1	16.8	6.5	2.3
2.8		87.7	42.9	19.0	7.5	2.7
2.9		93.8	46.8	21.2	8.7	3.2
3			50.9	23.6	9.9	3.8
1.1			55.2	26.2	11.3	4.4
3.2			59.6	28.9	12.7	5.1
3.3			64.2	31.7	14.3	5.9
3.4			68.9	34.7	16.0	6.7
3.5			73.8	37.8	17.7	7.6
3.6			78.8	41.0	19.7	8.6
1.7			83.9	44.4	21.7	9.7
3.8			89.1	48.0	23.8	10.9
3.9			94.5	51.6	26.1	12.2
4				55.4	28.5	13.5
4.1				59.3	31.0	15.0
1.2				63.4	33.6	16.5
1.3				67.5	36.3	18.1
4.4				71.8	39.2	19.9
4.5				76.2	42.2	21.7
4.6				80.8	45.3	23.7
4.7				85.4	48.5	25.7
4.8				90.2	51.8	27.8
4.9				95.0	55.2	30.1
5					58.8	32.4
5.1					62.4	34.9
5.2					66.2	37.4
5.3					70.0	40.0
5.4					74.0	42.8
5.5					78.1	45.6
5.6					82.3	48.6
5.7					86.6	51.6
5.8					90.9	54.8
5.9					95.4	58.0
6						61.4
5.1						64.8
5.2						68.4
6.3						72.0
6.4						75.7
6.5						79.5

Appendix B

QUEUE LENGTH (Lq)

C/S	Number of Units					
	2	3	4	5	6	7
1	0.33	0.05	0.01	0.00	0.00	0.00
1.1	0.48	0.07	0.01	0.00	0.00	0.00
1.2	0.68	0.09	0.02	0.00	0.00	0.00
1.3	0.95	0.13	0.02	0.00	0.00	0.00
1.4	1.35	0.18	0.03	0.01	0.00	0.00
1.5	1.03	0.24	0.04	0.01	0.00	0.00
1.6	2.84	0.31	0.06	0.01	0.00	0.00
1.7	4.43	0.41	0.08	0.02	0.00	0.00
1.8	7.67	0.53	0.11	0.02	0.00	0.00
1.9	17.59	0.69	0.14	0.03	0.01	0.00
2		0.89	0.17	0.04	0.01	0.00
2.1		1.15	0.22	0.05	0.01	0.00
2.2		1.49	0.28	0.07	0.02	0.00
2.3		1.95	0.35	0.08	0.02	0.00
2.4		2.59	0.43	0.10	0.03	0.01
2.5		3.51	0.53	0.13	0.03	0.01
2.6		4.93	0.66	0.16	0.04	0.01
2.7		7.35	0.81	0.20	0.05	0.01
2.8		12.27	1.00	0.24	0.07	0.02
2.9		27.19	1.23	0.29	0.08	0.02
3			1.53	0.35	0.10	0.03
3.1			1.90	0.43	0.12	0.03
3.2			2.39	0.51	0.15	0.04
3.3			3.03	0.62	0.17	0.05
3.4			3.91	0.74	0.21	0.06
3.5			5.17	0.88	0.25	0.08
3.6			7.09	1.06	0.29	0.09
3.7			10.35	1.26	0.35	0.11
3.8			16.94	1.52	0.41	0.13
3.9			36.86	1.83	0.48	0.15
4				2.22	0.57	0.18
4.1				2.70	0.67	0.21
4.2				3.33	0.78	0.25
4.3				4.15	0.92	0.29
4.4				5.27	1.08	0.34
4.5				6.86	1.26	0.39
4.6				9.29	1.49	0.45
4.7				13.38	1.75	0.53
4.8				21.64	2.07	0.61
4.9				46.57	2.46	0.70
5					2.94	0.81
5.1					3.54	0.94
5.2					4.30	1.08
5.3					5.30	1.25
5.4					6.66	1.44
5.5					8.59	1.67
5.6					11.52	1.94
5.7					16.45	2.26
5.8					26.37	2.65
5.9					56.30	3.11
6						3.68
6.1						4.39
6.2						5.30
6.3						6.48
6.4						8.08
6.5						10.34

ANSWERS TO PROBLEMS[14]

Chapter 2

1. Fires by Hour of Day - Boston 1988

Midnight - 4 a.m.	1531
4 a.m. - 8 a.m.	580
8 a.m. - Noon	772
Noon - 4 p.m.	1367
4 p.m. - 8 p.m.	1856
8 p.m. - Midnight	2219

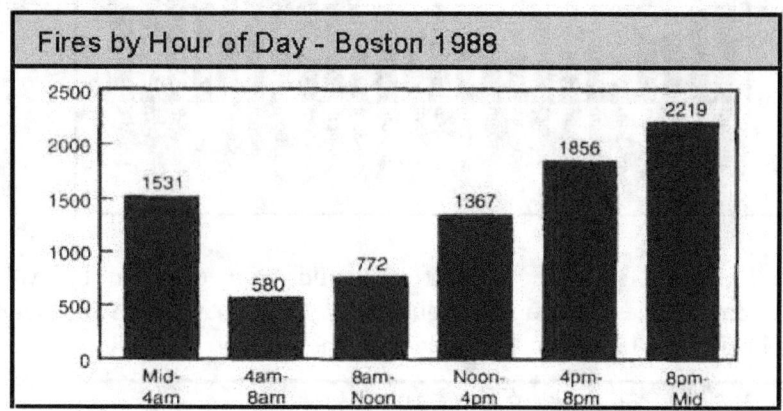

This histogram can be used to easily distinguish the morning hours from the afternoon and evening hours. The histogram in Exhibit 2-2 would take a little more concentration. This histogram would be useful when deciding shift assignments. The histogram in Exhibit 2-2 quickly shows which hour of day is the busiest and which hour is the least busiest. Remember, you always lose information when you aggregate data.

2. Exhibit 2-2 indicates that we should schedule more firefighters in the afternoon and evening hours and fewer firefighters in the early morning hours.

Exhibit 2-3 indicates that we need more firefighters on the weekend and not as many during the middle of the week.

---

14. Please note, the answers to the statistical problems were calculated by a scientific calculator. Your answers might not exactly match the answers that we have provided.

Exhibit 2-4 indicates that we should schedule more firefighters in the summer months (June and July). We do not need as many in January and February.

3. The distribution of fires by hour of day is almost identical to the 1988 data. There are more fires in the afternoon and evening hours, and very few fires in the early morning hours.

8-9pm was the busiest hour in 1990 and 9-10pm was the busiest hour in 1988.

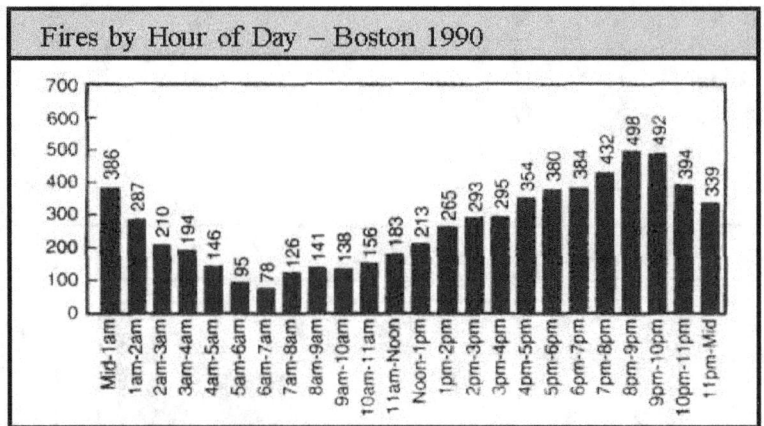

Like Exhibit 2-3, more firefighters would be needed on the weekend. Tuesdays have been a little busier in 1990 than in 1988. Sunday was the busiest in 1990, but Saturday was the busiest in 1988.

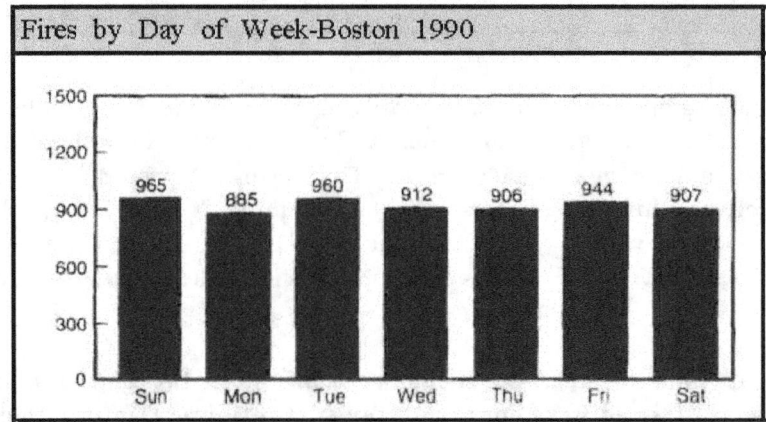

The 1990 data also indicates that there are more fires in the summer months (June and July). July was the busiest in 1990 and June was the busiest in 1988.

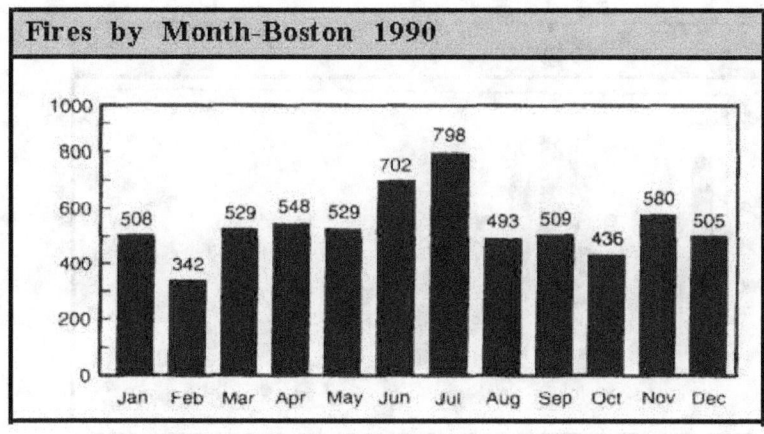

Fires by Month-Boston 1990

4. a. Jersey City, New Jersey, 1988: Ages of Civilian Casualties

Age Group	Frequency	Cumulative Frequency	Cumulative Percent
1-5	14	14	16.9
6-10	12	26	31.3
11-15	7	33	39.7
16-20	1	34	41.0
21-25	7	41	49.4
26-30	10	51	61.4
31-35	5	56	67.5
36-40	4	60	72.3
41-45	4	64	77.1
46-50	2	66	79.5
51-55	3	69	83.1
56-60	4	73	87.9
61-65	2	75	90.4
66-70	3	78	94.0
71-75	3	81	97.6
76-80	1	82	98.8
81-85	0	82	98.8
86-90	0	82	98.8
91-95	1	83	100.0
Total	83		

b. 39.7% of the civilian casualties were under 16 years of age.

c. 20.5% of the civilian casualties were over 50 years old.
(100.0 - 79.5=20.5)

5. The spikes in this histogram occur at <5 and 26-30. Exhibit 2-5 also has spikes at these age groups.

Answers: Chapter 2

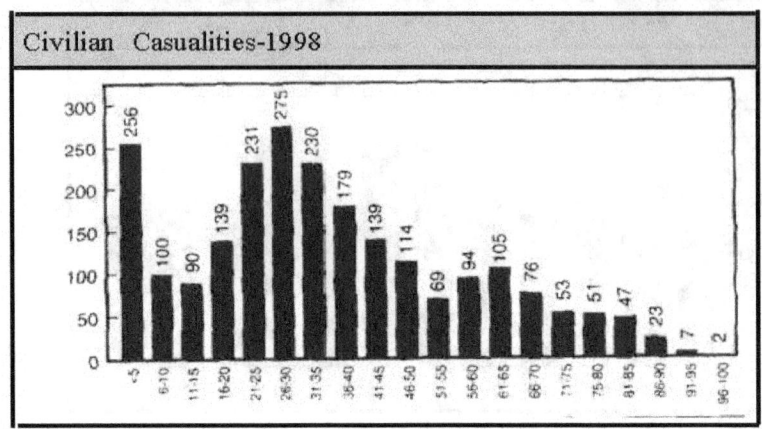

Civilian Casualities-1998

There is no hole or outlier in this group like there was in the Jersey City data.

6. There is no real pattern in this chart. The only thing you can tell is that there are more people killed or injured in the early morning hours.

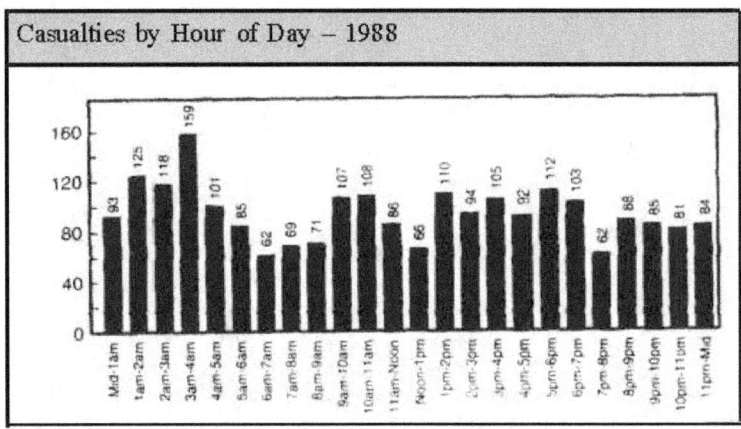

Casualties by Hour of Day – 1988

7. Peaks cause a cumulative distribution to increase.

Holes cause a cumulative distribution to increase slightly.

Spikes cause a cumulative distribution to increase sharply.

Gaps do not make the cumulative distributions increase. A cumulative distribution is always flat when a gap occurs.

Chapter 3

1.

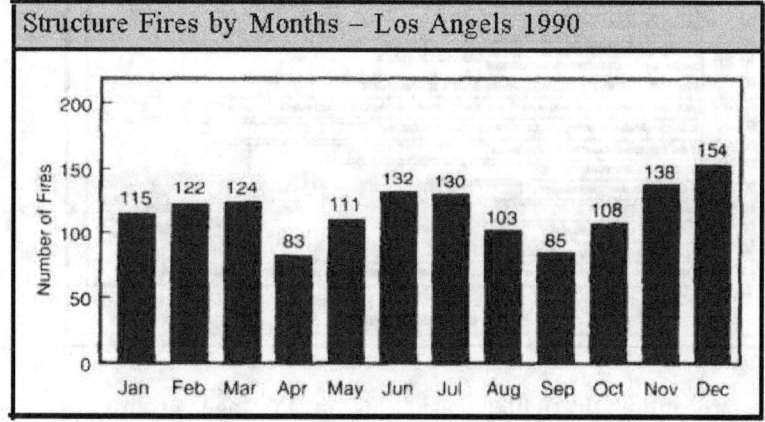

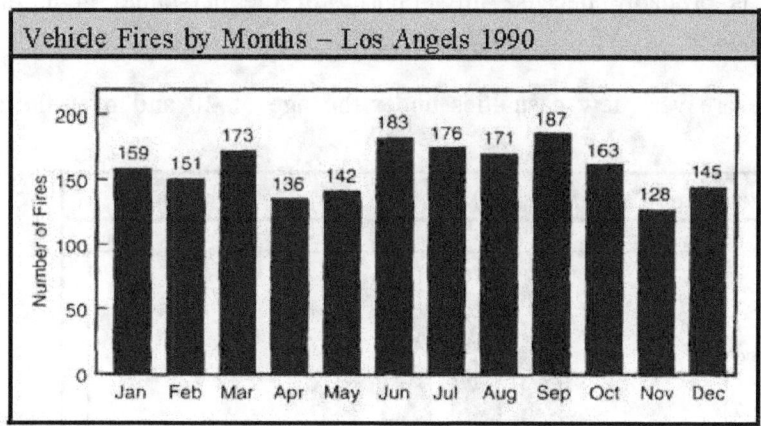

Overall there are more vehicle fires than structure fires. June and September were the months with the highest number of vehicle fires. December had the most structure fires. April and September had the lowest number of structure fires.

2. There are more male casualties than female casualties in all age groups except Over 65.

    a. No, because there are too many age groups. You should only use a pie chart to show how components relate to the whole and there should be no more than six components.

    b. By separating the data by sex, you can compare female and male casualties for each age group.

Answers: Chapter 3

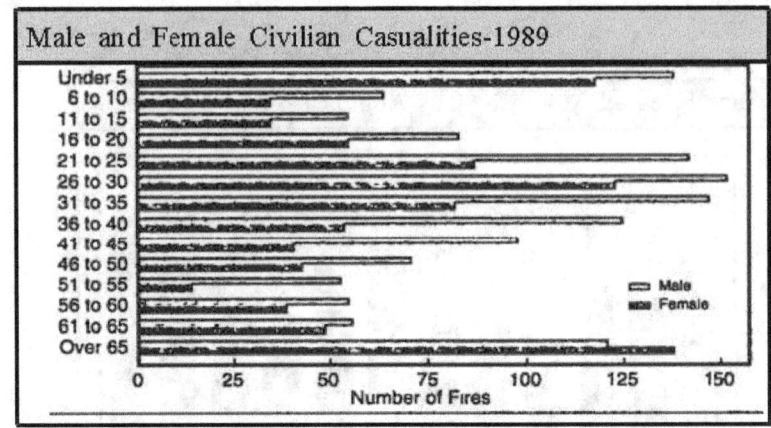

Male and Female Civilian Casualities-1989

3. a. There are more firefighter casualties in the age groups of 26-40. This is probably because the majority of the personnel is in these age groups.

There are very few casualties under the age of 20 and over the age of 55.

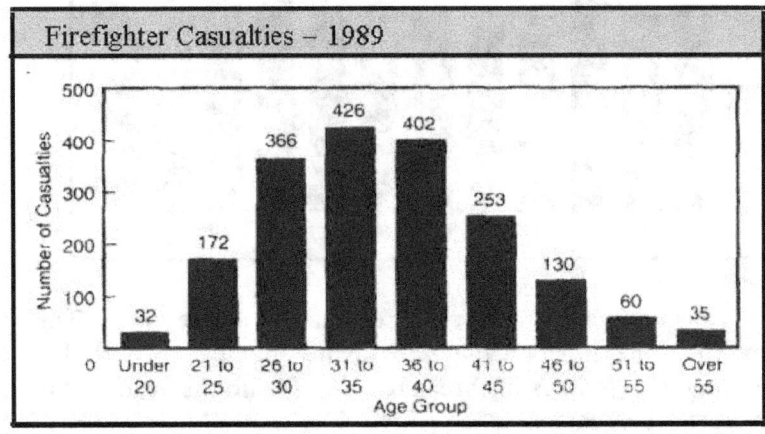

Firefighter Casualties – 1989

b.

Age Group	Number	Percent
Under 30	570	30.4%
31 to 40	828	44.1
41 to 50	383	20.4
Over 51	95	5.1

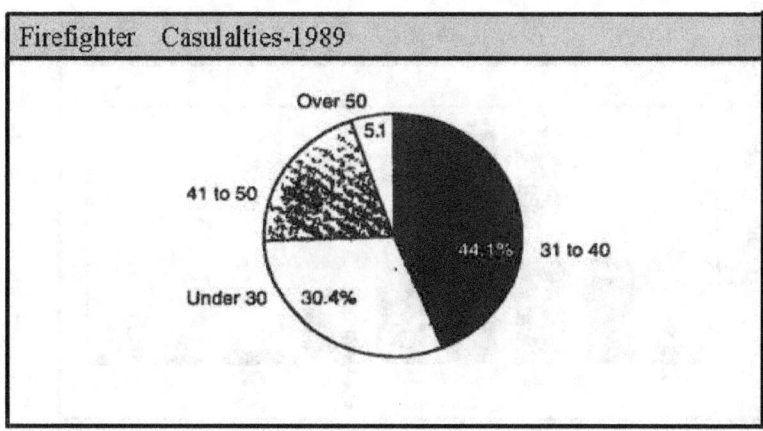

Firefighter   Casulalties-1989

4. Trees, brush, and grass fires make up a small percentage of fires in Chicago and Detroit.

Structure fires are predominant in Detroit where refuse fires are predominant in Chicago.

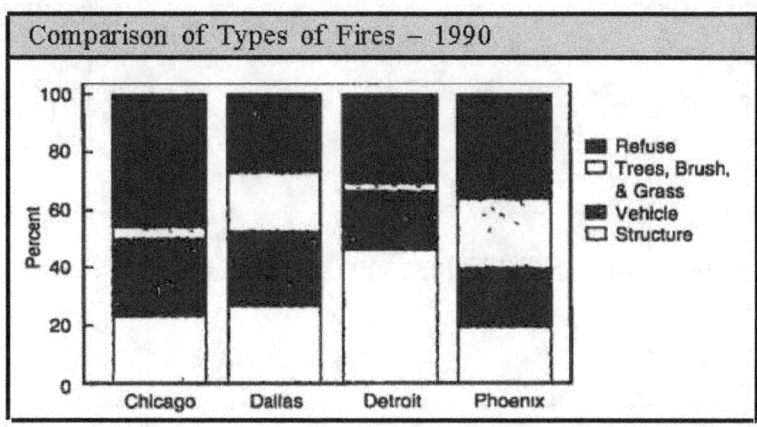

Comparison of Types of Fires – 1990

5. The two cities have similar distributions by Day of Week.

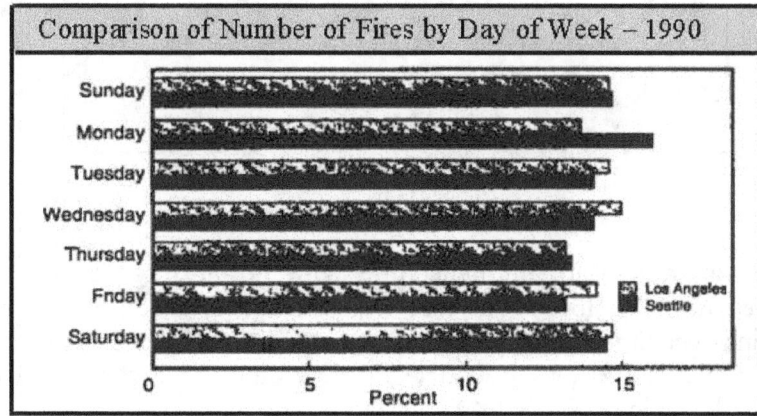

Comparison of Number of Fires by Day of Week – 1990

Answers: Chapter 3

6.

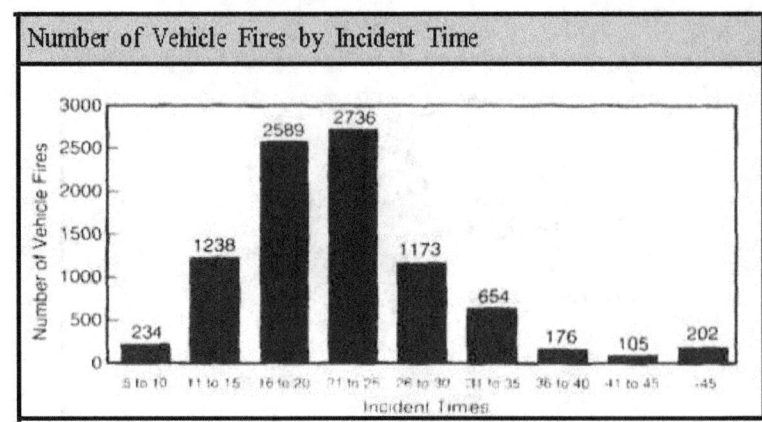

Number of Vehicle Fires by Incident Time

7. a.

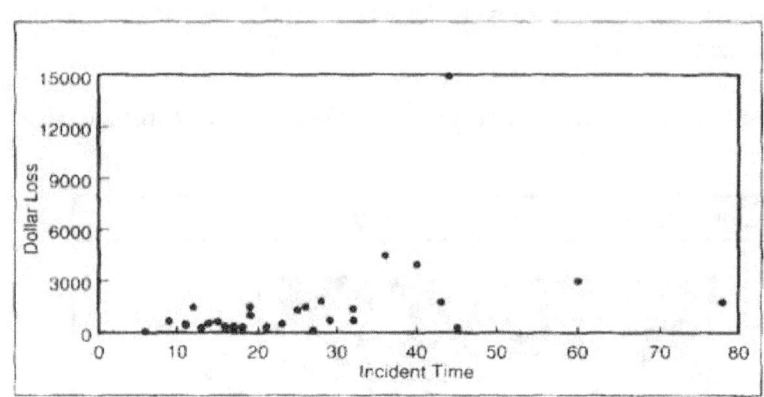

h. The outlier is 4-I minutes with $15000 for dollar loss.

c.

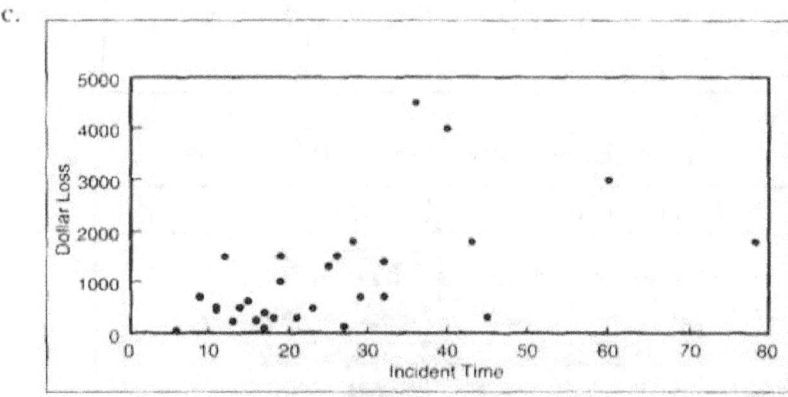

d. You can not relate incident time with dollar loss. You would expect the dollar loss to increase with incident time (especially when the incident time is over 30 minutes). As you can see in the scatter plot this is not always true. The incident time of 78 minutes had a lower dollar loss than 36, 40, and 60 minutes.

1. a.

	Travel Time	On-Scene Time	Dollar Loss
Mean	3.89	62.41	4923.67
Median	4	48	1000
Interquartile Range	2	66	4700

- To calculate the mean, sum all the data values and divide by the total number of data values.
- To determine the median, reorder each group of data and take the middle value (also known as the 50th percentile.

	Travel Time New Order	On-Scene Time New Order	Dollar Loss New Order	Cumulative Percent
	1	7	50	3.7
	1	9	100	7.4
	3	10	100	11.1
	3	10	150	14.8
	3	11	189	18.5
	3	20	250	22.2
25%	3	22	300	25.9
	3	23	300	29.6
	4	26	500	33.3
	4	27	500	37.0
	4	33	1000	40.7
	4	33	1000	44.4
	4	35	1000	48.1
50%	4	48	1000	51.9
	4	62	1500	55.6
	4	69	2000	59.2
	4	74	2000	63.0
	4	74	3000	66.7
	4	83	3000	70.4
	4	85	5000	74.1
75%	5	88	5000	77.8
	5	92	7000	81.5
	5	94	10000	85.2
	5	98	15000	88.9
	5	112	20000	92.6
	6	113	23000	96.3
	6	327	30000	100.0

- To determine the interquartile range, find the 25th and 75th percentiles and take the difference. To find the 25th and 75th percentiles, calculate the cumulative percent and pick the value whose cumulative percent is closest to (but not below) the percentile.

  Note: Since we reordered the data and listed all 27 values, the cumulative percent is the same for each of the variables.

b.

	Travel Time	Dollar Loss
Variance	1.41	62750996.1

c. Standard Deviation is the square root of the variance.

	Dollar Loss
Mean	4923.67
Standard Deviation	7021.553

One standard deviation about the mean is from -2997.883 to 12845.243 (mean - one standard deviation) to (mean + one standard deviation). There are 23 (85%) within one standard deviation.

Two standard deviations about the mean is from -10919.43 to 20766.776. There are 25 (93%) within two standard deviations.

d) For Dollar Loss, the large variance means that the spread of data around the mean is very large. For Travel Time, the small variance means that the spread around the mean is small.

e) Large dollar losses increase the mean, but usually have little effect on the median.

2.

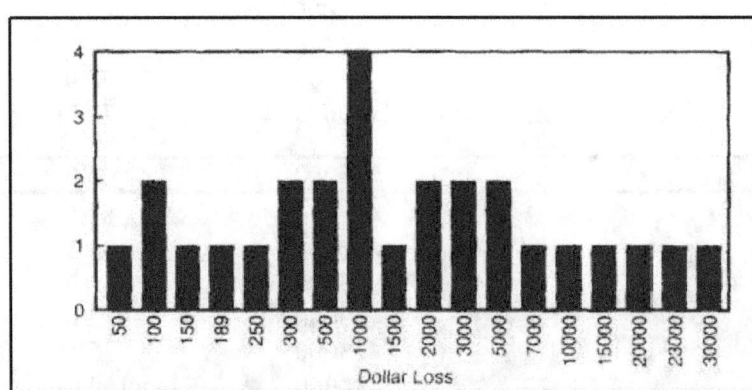

	Dollar Loss	Frequency	Cumulative Distribution	Cumulative Percent
	50	1	1	3.7
10%	**100**	**2**	**3**	**11.1**
	150	1	4	14.8
	189	1	5	18.5
	250	1	6	22.2
25%	**300**	**2**	**8**	**29.6**
	500	2	10	37.0
50%	**1000**	**4**	**14**	**51.9**
	1500	1	15	55.6
	2000	2	17	63.0
	3000	2	19	70.3
75%	**5000**	**2**	**21**	**77.8**
	7000	1	22	81.5
	10000	1	23	85.2
	15000	1	24	88.9
90%	**20000**	**1**	**25**	**92.6**
	23000	1	26	96.3
	30000	1	27	100.0

3. a.

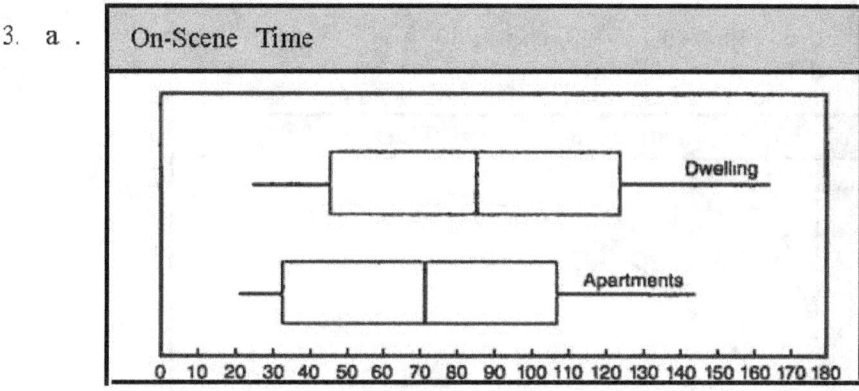

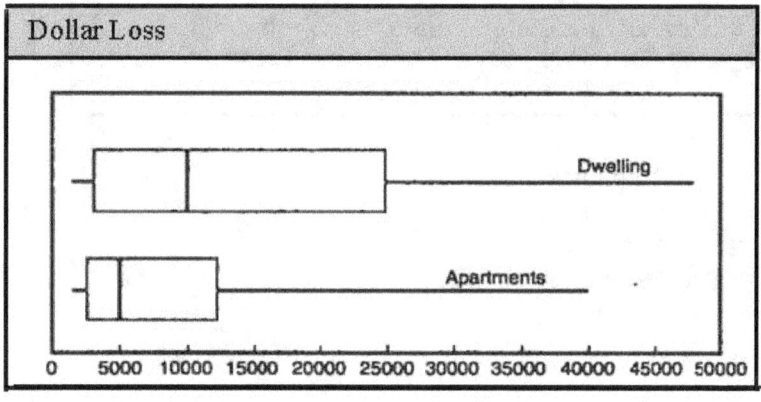

174   b. Interquartile Ranges

	On-Scene Time	Dollar Loss
Dwelling	78	22000
Apartments	75	9900

c. Fires in one- and two- family dwellings tend to have longer on-site times and larger dollar losses. For on-site times, the spread of the data is about the same for dwellings and apartments. The interquartile range is 78 for dwellings (123-45) and 75 for apartments (107-32). The dollar losses for dwellings have a greater spread than for apartments. The interquartile range is $22,000 for dwellings compared to $9,900 for apartments.

d. We would expect a greater variance with dwellings because of the larger spread of data.

4. a.

On-Scene Time

$$CS = \frac{3(62.41 - 48)}{63.36} = .682$$

Dollar Loss

$$CS = \frac{3(4923.67 - 1000)}{7921.553} = 1.486$$

b. CS will be zero when the mean and median are the same.

5. a. Drop 50, the 1st 100, 23000, and 30000.
Trimmed Mean = 3469.087

6. a.

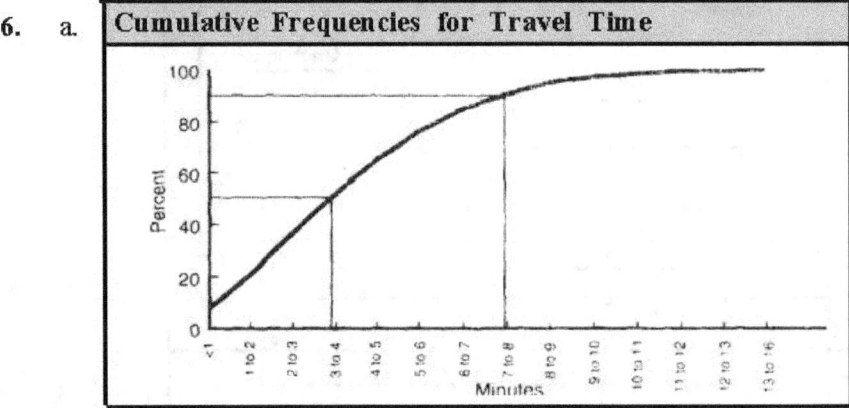

10% = 1 to 2 minutes
25% = 2 to 3 minutes
50% = 3 to 4 minutes
75% = 5 to 6 minutes
90% = 7 to 8 minutes

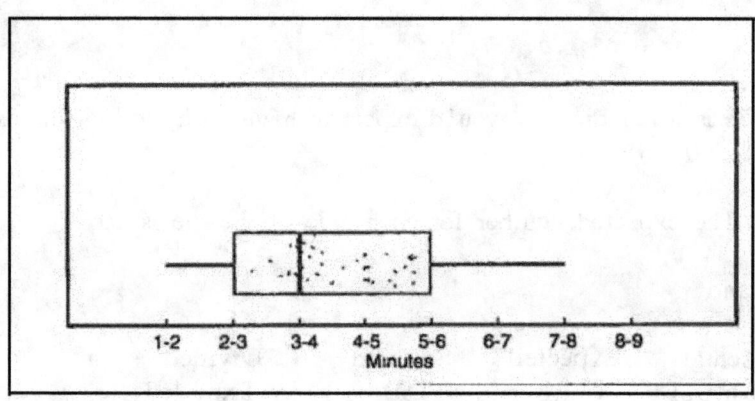

b. 90%

c. 3.75 minutes

1. If the die is a fair die, we would expect to have each side visible exactly 10 times (60 ÷ 6).

Step 1. The expected number for each side of the die is 10.

Step 2 and 3.

Dots Visible	Actual Number	Expected Number	Squared Diff.	Divided by Expected
One	13	10	9	.9
Two	11	10	4	.4
Three	7	10	9	.9
Four	12	10	4	.4
Five	9	10	1	.1
Six	7	10	9	.9

Step 4. Chi-squared statistic = sum of the results of the last column = 3.6

Step 5. Degrees of Freedom = $n - 1 = 5$

Step 6. Critical Chi-squared value = 11.07 (look in Appendix A and select the entry associated with 5 degrees of freedom)

2. The chi-squared statistic is less than the critical chi-squared value (3.6<11.07) so we determine that the die is a "fair" die at the 5 percent level.)

	Actual Number	Expected Number	Squared Diff.	Divided by Expected
Jan	467	355.67	12394.37	34.85
Feb	291	355.67	4182.21	11.76
Mar	392	355.67	13 19.87	3.71
Apr	322	355.67	1133.67	3.19
May	319	355.67	1344.69	3.78
Jun	349	355.67	44.89	.13
Jul	384	355.67	802.59	2.26
Aug	374	355.67	335.99	.94
Sep	359	355.67	11.09	.03
Oct	368	355.67	152.03	.43
Nov	298	355.67	3325.83	9.35
Dec	348	355.67	58.83	.16

Chi-squared Statistic = 70.59
Degrees of Freedom = 11
Critical Chi-squared Value = 19.68

a. Chi-squared statistic is greater than the critical chi-squared value. This means that the number of fires by month differs significantly from an equal distribution.

b. January, February, and November

3. a.

Type of Fire	Rest of U.S.	Percent	Metros	Percent
Structure	241,481	32.7	54,189	26.9
Outside of Structure	23,986	3.2	4,378	2.2
Vehicle	174,398	23.6	51,774	25.7
Trees, brush, grass	172,217	23.3	28,440	14.1
Refuse	111,291	15.1	59,767	29.7
Other	15,353	2.1	2,833	1.4
Total	738,726	100.00	201,381	100.00

b. Our null hypothesis is that we suspect that the distribution of fires in the Metro Areas does not differ significantly from the rest of the U.S.

Type of Fire	Actual Metros	Expected	Squared Diff.	Divided by Expected
Structure	54,189	65,851.59	136,016,005.51	2,065.49
Outside of Structure	4,378	6,444.19	4,269,141.12	662.48
Vehicle	51,744	47,525.92	17,792,198.89	374.37
Trees, brush, grass	28,440	46,921.77	341,575,822.33	7,279.69
Refuse	59,767	30,408.53	861,919,760.74	28,344.67
Other	2,833	4,229.00	1,948,816.00	460.82
			Calculated chi-squared statistic	39,187.52

Note: To calculate the expected, apply the percentages from the rest of the U.S. For example, 32.7% of the fires for the rest of the U.S. are structure fires. This means we expect that 32.7% of the Metro fires to be structure fires (65,851.59).

Calculated chi-squared statistic = 39,187.52
Degrees of Freedom = 5
Critical chi-squared value = 11.07
Our calculated chi-squared statistic is greater than the critical value.

Our conclusion is that the distribution of fires in the Metro Areas differs significantly from the rest of the U.S.

Answers: Chapter 5

c. Refuse fires amount for most of the differences in the two distributions. Structure fires and trees, brush, and grass fires also contribute significantly to the chi-squared Statistic.

4.

	Actual	Expected	Squared Diff.	Divided by Expected
Sunday	674	609.86	4,113.94	6.75
Monday	688	623.02	4,222.40	6.78
Tuesday	565	592.46	754.05	1.27
Wednesday	550	613.26	4,001.83	6.53
Thursday	576	585.25	85.56	.15
Friday	588	589.07	1.14	.002
Saturday	603	631.51	812.82	1.29
Total	4,244			22.77

Note: To calculate the expected. apply the percentages from the fires in the U.S. For example, 14.37% of the fires in the U.S. occurred on Sunday. This means we expect 14.37% of the fires to occur on Sunday in Denver. (14.37% times 4244 = 609.86).

Chi-squared statistic = 22.77
Degrees of Freedom = 6
Critical chi-squared value = 12.59

Since our calculated chi-squared statistic is greater than the critical value, our conclusion is that the distribution of fires by day of week for Denver differs from the rest of the country.

5.

	Actual	Expected	Squared Diff.	Divided by Expected
Sunday	674	644.24	885.66	1.37
Monday	688	634.90	2,819.61	4.44
Tuesday	565	596.28	978.44	1.64
Wednesday	550	598.40	2,342.-56	3.91
Thursday	576	570.39	31.47	.055
Friday	588	575.06	167.44	.29
Saturday	603	624.72	471.76	.755
Total	4.2441			12.46

Note: To calculate the expected, apply the percentages from the fires in the other metro cities. For example, 15.18% of the fires in the other metro cities occurred on Sunday-. This means we expect 15.18% of the fires to occur on Sunday,. in Denver. (15.18% times 4244 = 644.24).

Chi-squared statistic = 12.46
Degrees of Freedom = 6
Critical chi-squared value = 12.59

Since our calculated chi-squared statistic is less than the critical value, we can say that the distribution of fires by day of week in Denver does not differ significantly from the other Metro cities.

1.

$$\frac{x_{11} x_{22}}{x_{12} x_{21}} = \frac{36 \times 32}{24 \times 48} = \frac{1552}{1552} = 1$$

Since the Odds Ratio is equal to 1, the two variables have complete independence.

2. 1ˢᵗ Table

$$\frac{601 \times 897}{1,884 \times 166} = \frac{539,097}{312,744} = 1.72$$

2nd Table

$$\frac{837 \times 851}{1,648 \times 212} = \frac{712,287}{349,376} = 2.04$$

3ʳᵈ Table

$$\frac{801 \times 694}{1,684 \times 369} = \frac{555,894}{621,369} = .89$$

3. In order to verify the model values! you need to sum the mean and term values and take the natural antilogarithm. The term values listed in Exhibit 6-11 are when all 4 variables are present (i, j, k, and 1 are all 1). If a variable is not present (indicated by 2) then the term value is the negative value of the one given in Exhibit 6-11. The easiest way to see this is to develop tables for each of the terms.

Term A	i=1	i=2
	s77	-.577

Term C	k=1	k=2
	-.484	.484

Term B	J=1	j=2
	-.678	.678

Term D	l=1	l=2
	-.591	.591

Term AB	i=1	i=2
j=1	.120	-.120
j=2	-.120	.120

Term BD	j=1	j=2
l=1	.084	-.084
l=2	-.084	.084

Term AD	i=1	i=2
I=1	.169	-.169
I=2	-.169	.169

Term CD	k=1	k=2
l=1	-.326	.326
l=2	.326	-.326

Term BC	j=1	j=2
k=1	.008	-.008
k=2	-.008	.008

Term BCD	j=1		j=2	
	k=1	k=2	k=1	k=2
l=1	-.084	.084	.084	-.084
l=2	.084	-.084	-.084	.084

Model (1,1,1,1) $= 4.807 + .577 + (-.678) + (-.484) + (-591)$
$\quad\quad\quad\quad + .120 + .169 + .008 + .084 + (-.326) + (-.084)$
$\quad\quad\quad\quad = 3.602$
anti log $(3.062) = 36.67$ 　　　　 Actual $= 37$

Model (1,1,2,2) $= 4.807 + .577 + (-.678) + .484 + 591$
$\quad\quad\quad\quad + .120 + (-.169) + (-.008) + (-.084) + (-.326) + (-.084)$
$\quad\quad\quad\quad = 5.23$
anti log (5.23) 　　 $= 186.79$ 　　　　 Actual $= 181$

Model (2,1,2,1) $= 4.807 + (-577) + (-.678) + .484 + (-S91)$
$\quad\quad\quad\quad + (-.120) + (-.169) + (-.008) + .084 + .326 + ,084$
$\quad\quad\quad\quad = 3.642$
anti log $(3.642) = 38.17$ 　　　　 Actual $= 3.5$

4. a. Look at all standardized values with magnitudes greater than 2.0 (ignoring the sign). All four main effects (A,B,C, and D) are important. Also the following interaction effects are important (in order of magnitude).

AC 　　(6.3)
AD 　　(5.2)
BC 　　(3.5)
CD 　　(3.4)
BD 　　(3.1)
ABC　(3.1)
AB 　　(2.5)

b. Since variable A is the response variable, we want to model how the other three variables relate to A. Since variables B, C, and D are the explanatory variables, any hierarchical models must include the term BCD and any of its higher-order terms (B, C, D, BC, BD, and CD)

Possible hierarchical models would be:

Model	Degrees of Freedom
AB/ACD/BCD	3
AD/ABC/BCD	3
AC/ABD/BCD	3
ACD/BCD	4
AB/AC/AD/BCD	4
AB/AC/BCD	5
AB/AD/BCD	5
AC/AD/BCD	5

5.   a. Models 1 and 2. Remember the Y^2 statistic follows a chi-squared distribution. Use Appendix A to test whether a model is less than the entry in Appendix A at the 5 percent level. If the Y^2 statistic is less than the entry in Appendix A, then we accept the model as a good fit to the data.

b. Model 2. When there are two or more models that provide a good fit to the data. select the one with the fewest number of terms.

c.

TABLE 1	Detector Performed	Detector Did Not Perform	Total
Confined to Room	586	1789	2375
Extended Beyond Room	166	867	1033
Total	752	2656	

TABLE 2	Fire Started in Functional Area	Fire Started in Area Non-Functional	Total
Confined to Room	1681	694	2375
Extended Beyond Room	585	448	1033
Total	2266	1142	

TABLE 3	Equipment Involved	No Equipment Involved	Total
Confined to Room	834	1541	2375
Extended Beyond Room	216	817	1033
Total	1050	2358	

d. Table 1

The odds that a fire was confined to the room when the detector performed is 3.53:1. $(x_{11}/x_{21} = 586/166)$.

The odds that a fire extended beyond the room when the detector did not perform is 2.06:1.

$$\frac{x_{11}x_{22}}{x_{12}x_{21}} = \frac{586 \times 867}{166 \times 1789} = 1.71$$

Table 2

The odds that a fire was confined to the room when the fire was started in a functional area is 2.87:1.

The odds that a fire extended beyond the room when the fire was started in a non-functional area is 1.55:1.

$$\frac{x_{11}x_{22}}{x_{12}x_{21}} = \frac{1681 \times 448}{585 \times 694} = 1.85$$

Table 3

The odds that a fire was confined to the room when equipment was involved is 3.86:1.

The odds that a fire extended beyond the room when no equipment was involved is 1.89:1.

$$\frac{x_{11}x_{22}}{x_{12}x_{21}} = \frac{834 \times 817}{216 \times 1541} = 2.05$$

1.   a.

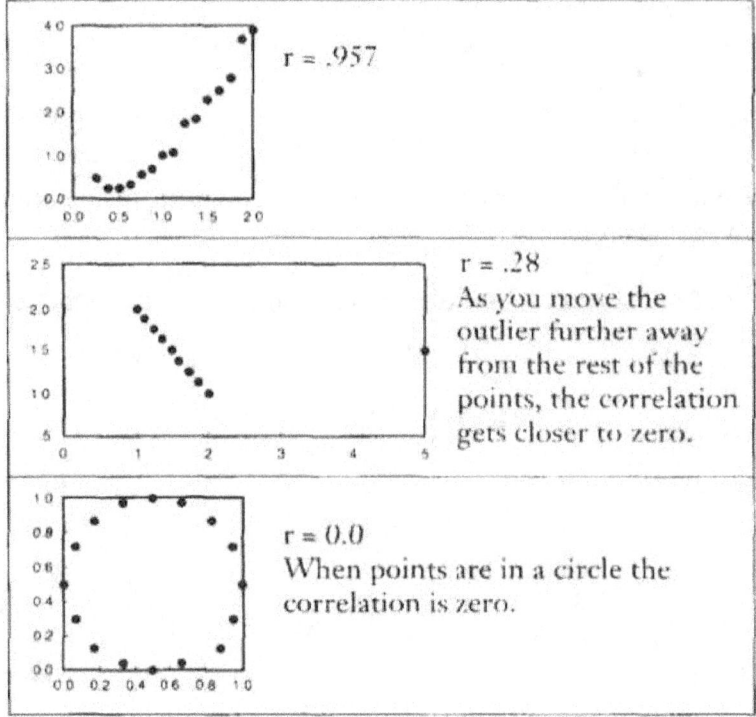

b.   r = 1

2.

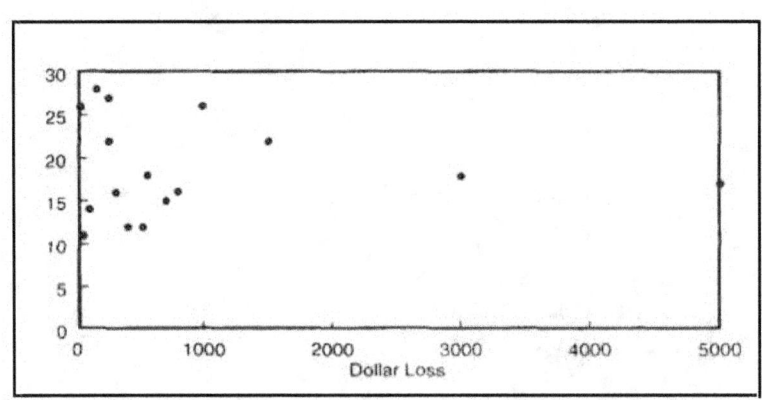

Correlation = -.065

Notes: 1) Calculate the average for each variable (sum of the values divided by the total number of values).

          2) Calculate the sample standard deviation.

$$\sqrt{\frac{\sum(x_i-\bar{x})^2}{n-1}}$$

3) Calculate the standard units for each value.

$$\frac{\text{Value - Average}}{\text{Standard Deviation}}$$

4) Calculate the product of the standard units for each pair.
5) Sum the resulting values and divide by the number of points minus 1.

$$\frac{-.98}{16 - 1} = -.065$$

Incident time	Dollar Loss	Incident Time Standard Units	Dollar Loss Standard Units	Product
15	700	-.658	-.221	.145
18	3000	-.132	1.524	-.201
22	250	.570	-.563	-.321
26	35	1.272	-.726	-.923
14	100	-.834	-.677	.565
12	500	.1.185	-.373	.442
28	150	1.623	-.639	-1.037
16	300	-.483	-.525	.254
12	400	.1.185	-.449	.532
16	800	-483	-.145	.070
26	1000	1.272	.006	.008
22	1500	.570	.386	.220
17	5000	-.307	3.042	-.934
18	550	-.132	-.335	.044
27	250	1.448	-.563	-.815
11	50	-1.360	-.714	.971

Average Incident Time = 18.75
Standard Deviation for Incident Time = 5.698
Average Dollar Loss = 991.563
Standard Deviation for Dollar Loss = 1317.791

3. a. Estimated Fire Loss = (206.2 x 45) -6266.4
(45 minutes)   = 3012.60

Estimated Fire Loss = (206.2 x 180) -6266.4
(180 minutes)   =30849.60

b. The slope = 206.2 means that the fire loss increases by 206.2 every time the incident time increases by 1 minute. When the incident time increases by 10 minutes, the fire loss will increase by 2062.00 (206.2 x 10).

Answers: Chapter 7

c. Incident time is from time of dispatch to time back in service. Sometimes a fire has been going on for a while before the fire department is called, thus resulting in high fire loss even though the incident time is under 30 minutes.

4. a.

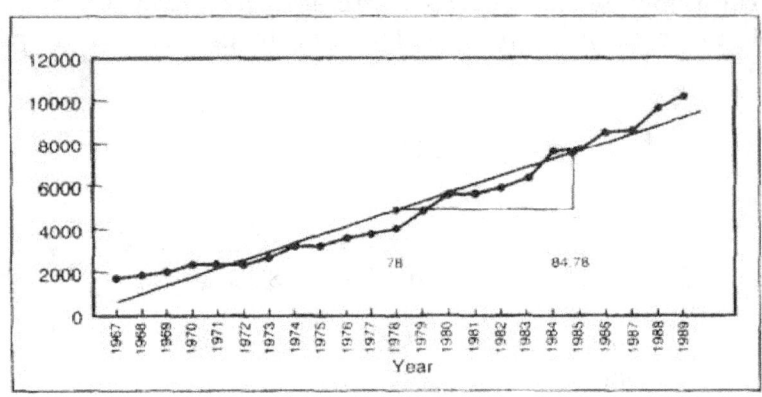

$$m = \frac{r x S.D. \text{ of Fire Loss}}{S.D. \text{ of Years}} \qquad \frac{.9753 \times 2696.394}{6.782} = 387.76$$

Determine the intercept:

Fire Loss Average = m x Year Average + b
4929.043 = (387.76 x 78) + b
-25316.237 = b

Regression Line is: Fire Loss = 387.76 x Year -25316.237

Note: If you have Fire Loss along the x-axis and year along the y-axis, your regression line will be: Year = .(0025 x Fire Loss + 65.68

c. Year   Estimated Fire Loss

1988   8806.64
1989   9194.40
1990   9582.16

Fire Loss = 387.76 x 88 -25316.237

d. We want the regression line to pass through one point that represents the two averages. Put another point by moving one standard deviation for the year to the right and upward by one standard deviation for Fire Loss times the correlation.

Chapter 8

1. Fires = 2.0 + (.0017 x POP) + (.068 x BOARDED) +
   (.013 x FAMTYPE) + (.060 x DENSITY)

   <u>Example</u>
   Census Tract 709

   Fires = 2.0 + (.0017 x 1386) + (.068 x 19) + (.013 x 184) + (.060 x 100)
   = 2.0 + 2.3562 + 1.292 + 2.392 + 6
   = 14.0402

Census Tract	Actual Fires	Estimated Fires
709	15	14.04
812	30	29.68
902	30	16.77
907	10	10.20

2. Travel Time = 5.27 + (1.38 x DELAYS) + (.36 x CALLTYPE) +
   (3.83 x AREA1) + (2.48 x AREA2)

   Remember: CALLTYPE   0 for ALS Calls
                             1 for BLS Calls

                DELAYS   0 not delayed
                             1 delayed

                AREA1   0 not Area 1
                             1 Area 1

                AREA2   0 not Area 2
                             1 Area 2

   a. Travel Time = 5.27 + (1.38 x 0) + (.36 x 1) + (3.83 x 1) + (2.48 x 0)

      = 9.46

   b. Travel Time = 5.27 + (1.38 x 1) + (.36 x 0) + (3.83 x 0) + (2.48 x 1)
      = 9.13

   c. Travel Time = 5.27 + (1.38 x 0) + (.36 x 0) + (3.83 x 0) + (2.48 x 0)
      = 5.27

   d. Travel Time = 5.27 + (1.38 x 1) + (.36 x 1) + (3.83 x 0) + (2.48 x 0)
      = 7.01

Answers: Chapter 8

1. a.

$$\text{Unit Utilization (3)} = \frac{3.5 \times 45}{3 \times 60} = \frac{157.5}{180} = .875 \qquad 87.5\%$$

Unit Utilization (4) = 65.6%
Unit Utilization (5) = 52.5%
Unit Utilization (6) = 43.7%

b. A change from 4 units to 5 units will result in a difference of 13.1%
A change from 5 units to 6 units will result in a difference of 8.8%

2. a. To obtain the Probability of Delay you have to:

1) Calculate the Table Key = $\frac{ct}{60}$ where c is the number of calls per hour and t is the average time per call.
2) Use Appendix B to find the probability that a call will be delayed. The number of units are across the top and the Table Key is shown down the left column.

$$\text{Table Key} = \frac{ct}{60} = \frac{3.5 \times 45}{60} = 2.625 \qquad 2.6 \text{ rounded}$$

Probability of Delay for:

3 Units = 75.9%
4 Units = 35.4%
5 Units = 14.9%
6 Units = 5.6%

b. To determine the average number of citizens waiting, use Appendix C. The number of units is displayed across the top and the Table Key is shown down the left column.

Average Number of Citizens Waiting:

3 Units = 4.93
4 Units = .66
5 Units = .16
6 Units = .04

c. Waiting Time $= \dfrac{\text{Appendix C Entry} \times 60}{c}$ where c is the number of calls per hour.

3 Units 84.51 minutes (4.93 times 60 divided by 3.5)
4 Units 11.31 minutes
5 Units 2.74 minutes
6 Units .69 minutes

3. a. 6% of 3.5 calls per hour = .21
       3.5 + .21 = 3.71
       3.71 calls per hour

    b.

    $$\text{Unit Utilization} = \frac{ct}{60n} = \frac{3.71 \times 50}{60 \times 4} = =.773 \quad 77.3\%$$

    $$\text{Table Key} = \frac{ct}{60} = \frac{3.71 \times 50}{60} = 3.09 = 3.1 \text{ rounded}$$

    Probability of Delay for 4 units is 55.2%
    Average Number People Waiting is 1.9

    C. In question #2, the probability of delay for 4 units was 35.4% and the average number of people waiting was .66. The 6% increase in calls per hour and the 5 minute increase on each call greatly increased the probability of delay and tripled the number of people who would be waiting.

4. There has to be an analysis of records available to your department. This could be a computer aided dispatch system (CAD) which would have good time information to determine how busy units are. You should also perform an analysis of schedules in order to determine how many units are fielded.

    If you are going to calculate unit utilization, you need to find out how many units are fielded and not scheduled. You might want to look at daily rosters or time cards to determine how many units were actually fielded.

    Also read over the section in chapter one about data quality.

5. There is no one way of setting objectives. There needs to be a decision made by the command staff of the fire department. It is also a good idea to involve someone from the city or county when setting the objectives. One popular approach is to perform an analysis of current operations to determine current performance levels. Then determine if the command staff is satisfied with current performance or wants to make improvements. Queuing analysis is a good first step in determining staffing needs for improvements.

6. a. Unit utilization not more than 60 percent   5 units
       Probability of delay not more than 3 percent   7 units
       Average number of citizens waiting not to exceed 25   5 units

Unit Utilization $= \dfrac{ct}{60n}$ $.60 = \dfrac{4 \times 38}{60n}$ $n = 4.2$

If you had 4 units, the unit utilization would be 63.3%
If you had 5 units, the unit utilization would be 50.67%

Table Key $= \dfrac{ct}{60} = \dfrac{4 \times 38}{60} = 2.53$

Probability of Delay for: Average Number of Citizens Waiting:

     4 units = 32%   4 units = .53
     5 units = 13%   5 units = .13
     6 units = 4.7%   0 units = .03
     7 units = 1.5%   7 units = .01

b. 7 units

Unit Utilization (7 units) $= \dfrac{4 \times 38}{m} = .362$ 36.2%

7. a. 4 units

Waiting Time $= \dfrac{\text{Appendix C Entry} \times 60}{c}$

Waiting Time (4 units) $= \dfrac{.53 \times 60}{4} = 7.05$ minutes

Waiting Time (5 units $= \dfrac{.13 \times 60}{4}$ 1 .05 minutes

			Expected Percentage
EMS Calls	Number of Hours	Percent	of EMS Calls
0	158	26.0	23.2
1	192	31.6	33.9
2	144	23.6	24.8
3	68	11.2	12.0
4	34	5.6	4.4
5	10	1.6	1.3
6	2	.3	.3
Total	608		

Note: Poisson Distribution: $P(x=k) = \dfrac{e^{-c} c^k}{k!}$

Where c is the average number of calls per hour and k is an integer value starting with 0.

$$\frac{e^{-1.46} 1.46^0}{0!} = \frac{.2322}{0!} = .232 \quad 23.2\%$$

b. If the calls follow a Poisson distribution, we should be able to obtain a straight line by plotting k against the quantity:

$$\log\left(\frac{k! \, n_k}{N}\right)$$

where N is the total number of hours.

Note: by definition $0! = 1$.

0 calls $\log\left(\dfrac{0! \times 158}{608}\right) = -1.35$

0 calls -1.35
1 call -1.15
2 calls -.75
3 calls -.40
4 calls .29
5 calls .68
6 calls .86

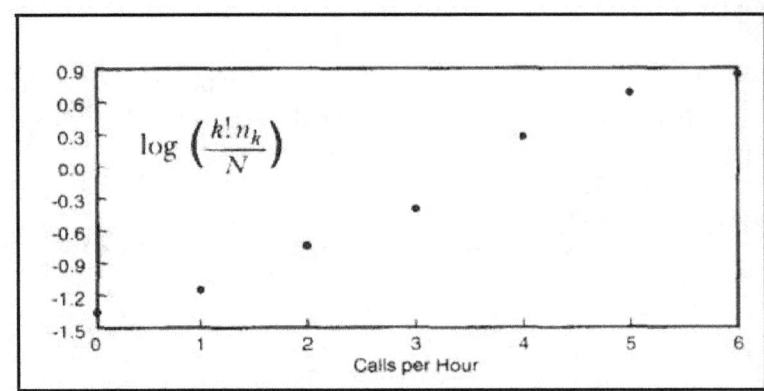

c. The conclusion is that the Poisson distribution gives a good approximation to the experienced distribution because it produces a relatively straight line.

9. a.

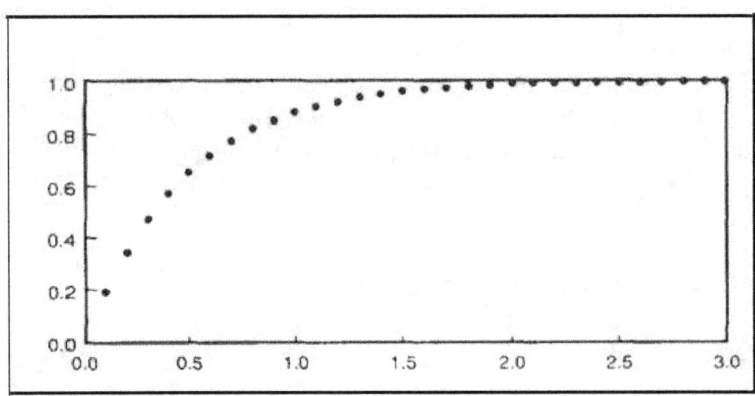

b. 25th percentile = .804
median = .962
75th percentile = .993

References

Agresti, Alan, and Barbara Finlay. *Statistical Methods for the Social Sciences.* San Francisco, California: Dellen Publishing Company, 1986.

Cleveland, William S. *The Elements of Graphing Data.* Monterey, California: Wadsworth Advanced Books and Software, 1985.

Devore, Jay and Roxy Peck. *Statistics: The Exploration and Analysis of Data.* New York, New York: West Publishing Company, 1986.

Freedman, David, Robert Pisani, and Roger Purves. *Statistics.* New York, New York: W.W. Norton & Company, 1978.

Goodman, L.A. "The Analysis of Multidimensional Contingency Tables, Stepwise Procedures and Direct Estimation Methods for Building Models for Multiple Classifications." *Technometrics* (1971): 13, 33-61.

Hoaglin, David D., Federick Mosteller, and John W. Tukey. *Exploring Data Tables, Trends, and Shapes.* New York, New York: John Wiley and Sons, 1985.

Jaffe, A. J. and Herbert F. Spirer. *Statistics: Straight Talk for Twisted Numbers.* New York, New York: Marcel Dekker, Inc., 1987.

Mosteller, Frederick, Stephen E. Fienberg, and Robert E.K. Rourke. *Beginning Statistics with Data Analysis.* Reading, Massachusetts: Addison-Wesley Publishing Company, 1983.

Phillips, John J., Jr. *How to Think About Statistics.* New York, New York: W.H. Freeman and Company, 1988.

Schefler, William C. *Statistics: Concepts and Applications.* Menlo Park, California: The Benjamin/Cummings Publishing Company, Inc., 1988.

Siegel, Andrew F. *Statistics and Data Analysis: An Introduction.* New York, New York: John Wiley & Sons, Inc., 1988.

Tanur, Judith M., Frederick Mosteller, William H. Kruskal, Erich L. Lehmann, Richard F. Link, Richard S. Pieters, and Gerald R. Rising. *Statistics: A Guide to the Unknown.* Pacific Grove, California: Wadsworth & Brooks/Cole Advanced Books & Software (Third Edition), 1988.

Upton, Graham J.G. *The Analysis of Cross-tabulated Data.* New York, New York: John Wiley & Sons, 1978.

Zeisel, Hans. *Say It With Figures.* New York, New York: Harper & Row Publishers, Inc., 1985.

Zelazny, Gene. *Say It With Charts.* Homewood, Illinois: Dow Jones-Irwin, Inc., 1985.

Key Equations

Sample Mean $= \bar{x} = \dfrac{\sum x_i}{n}$

Regression Line $= y = mx + b$

Sample Variance $= s^2 = \dfrac{\sum (x_i - \bar{x})^2}{n-1}$

Standard Error $= \sqrt{\dfrac{SSE}{n-k-1}}$

where k = number of independent variables

Sample Standard Deviation $= s = \sqrt{\dfrac{\sum (x_i - \bar{x})^2}{n-1}}$

Sum of Squared Errors $= SSE = \sum (y_i - \hat{y}_i)^2$

Chi-squared Value $= \sum \dfrac{(\text{observed} - \text{expected})^2}{\text{expected}}$

Total Sum of Squares $= SST = \sum (y_i - \bar{y})^2$

Expected Value $= \dfrac{\text{Row Sum} \times \text{Column Sum}}{\text{Grand Total}}$

Coefficient of Determination $= R^2 = \dfrac{SST - SSE}{SST}$

Odds Ratio $= \dfrac{x_{11}\, x_{22}}{x_{12}\, x_{21}}$

Unit Utilization $= \dfrac{ct}{60n}$

Standard Units $= \dfrac{x_i - \bar{x}}{s}$

Queuing Table Key $= \dfrac{ct}{60}$

Correlation $= r = \dfrac{\sum (x_i - \bar{x})(y_i - \bar{y})}{(n-1)\, s_x s_y}$

Waiting Time $= \dfrac{\text{Queue Length} \times 60}{c}$

Slope $= m = \dfrac{r \times s_x}{s_y}$

Poisson Distribution $= P(x = k) = \dfrac{e^{-c} c^k}{k!}$

Intercept $= b = \bar{y} - m\bar{x}$

Exponential Distribution $= f(x) = u e^{-ux}$